CRISPR–Cas

A LABORATORY MANUAL

ALSO FROM COLD SPRING HARBOR LABORATORY PRESS

RELATED TITLES

Decoding the Language of Genetics

DNA Recombination

DNA Repair, Mutagenesis, and Other Responses to DNA Damage

Introduction to Protein–DNA Interactions

Microbial Evolution

Quickstart Molecular Biology

RNA: Life's Indispensable Molecule

RNA Worlds: From Life's Origins to Diversity in Gene Regulation

OTHER LABORATORY MANUALS

Antibodies: A Laboratory Manual, Second Edition

Budding Yeast: A Laboratory Manual

Calcium Techniques: A Laboratory Manual

Cell Death Techniques: A Laboratory Manual

Manipulating the Mouse Embryo: A Laboratory Manual, Fourth Edition

Molecular Cloning: A Laboratory Manual, Fourth Edition

Molecular Neuroscience: A Laboratory Manual

Mouse Models of Cancer: A Laboratory Manual

Purifying and Culturing Neural Cells: A Laboratory Manual

RNA: A Laboratory Manual

Subcellular Fractionation: A Laboratory Manual

HANDBOOKS

A Bioinformatics Guide for Molecular Biologists

At the Bench: A Laboratory Navigator, Updated Edition

At the Helm: Leading Your Laboratory, Second Edition

Career Options for Biomedical Scientists

Experimental Design for Biologists, Second Edition

Lab Math: A Handbook of Measurements, Calculations, and Other Quantitative Skills for Use at the Bench

Lab Ref: A Handbook of Recipes, Reagents, and Other Reference Tools for Use at the Bench, Volume 1 and Volume 2

Next-Generation DNA Sequencing Informatics, Second Edition

Statistics at the Bench: A Step-by-Step Handbook for Biologists

Using R at the Bench: Step-by-Step Data Analytics for Biologists

WEBSITE

www.cshprotocols.org

CRISPR–Cas

A LABORATORY MANUAL

EDITED BY

Jennifer Doudna
University of California, Berkeley

Prashant Mali
University of California, San Diego

CSH PRESS
COLD SPRING HARBOR LABORATORY PRESS
Cold Spring Harbor, New York • www.cshlpress.org

CRISPR–Cas
A LABORATORY MANUAL

All rights reserved
© 2016 by Cold Spring Harbor Laboratory Press, Cold Spring Harbor, New York
Printed in the United States of America

Publisher	John Inglis
Acquisition Editor	Richard Sever
Managing Editor	Maria Smit
Director of Editorial Services	Jan Argentine
Project Manager	Maryliz Dickerson
Permissions Coordinator	Carol Brown
Production Editor	Joanne McFadden
Production Manager	Denise Weiss
Director of Product Development & Marketing	Wayne Manos
Cover Designer	Denise Weiss

Cover art: CRISPR–Cas adaptive immunity in bacteria and archaea relies on the capture of invading foreign DNA and its subsequent integration into the CRISPR locus to elicit memory of the infection. The RNA products of newly captured DNA, called spacers, are used by Cas proteins, such as Cas9, as guides for destruction of complementary foreign DNA sequences. The image depicts the process of capturing foreign DNA, catalyzed by the proteins Cas1 and Cas2, in preparation for integration into the host's CRISPR locus. Credit: Megan Riel-Mehan and Graham Johnson (UCSF).

Library of Congress Cataloging-in-Publication Data

Names: Doudna, Jennifer A., editor. | Mali, Prashant, editor.
Title: CRISPR-Cas: a laboratory manual/edited by Jennifer Doudna,
Prashant
Mali.
Description: Cold Spring Harbor, New York: Cold Spring Harbor Laboratory
Press, [2016] | Includes bibliographical references and index.
Identifiers: LCCN 2015050261 | ISBN 9781621821304 (hardcover: alk.
paper) |
ISBN 9781621821311 (pbk. : alk. paper)
Subjects: | MESH: CRISPR-Cas Systems | CRISPR-Associated Proteins |
Genetic
Techniques | Laboratory Manuals
Classification: LCC QP623 | NLM QU 25 | DDC 572.8/8078--dc23
LC record available at http://lccn.loc.gov/2015050261

Students and researchers using the procedures in this manual do so at their own risk. Cold Spring Harbor Laboratory makes no representations or warranties with respect to the material set forth in this manual and has no liability in connection with the use of these materials. All registered trademarks, trade names, and brand names mentioned in this book are the property of the respective owners. Readers should please consult individual manufacturers and other resources for current and specific product information.

With the exception of those suppliers listed in the text with their addresses, all suppliers mentioned in this manual can be found on the BioSupplyNet website at www.biosupplynet.com.

All World Wide Web addresses are accurate to the best of our knowledge at the time of printing.

Procedures for the humane treatment of animals must be observed at all times. Check with the local animal facility for guidelines.

Certain experimental procedures in this manual may be the subject of national or local legislation or agency restrictions. Users of this manual are responsible for obtaining the relevant permissions, certificates, or licenses in these cases. Neither the authors of this manual nor Cold Spring Harbor Laboratory assume any responsibility for failure of a user to do so.

The materials and methods in this manual may infringe the patent and proprietary rights of other individuals, companies, or organizations. Users of this manual are responsible for obtaining any licenses necessary to use such materials and to practice such methods. COLD SPRING HARBOR LABORATORY MAKES NO WARRANTY OR REPRESENTATION THAT USE OF THE INFORMATION IN THIS MANUAL WILL NOT INFRINGE ANY PATENT OR OTHER PROPRIETARY RIGHT.

Authorization to photocopy items for internal or personal use, or the internal or personal use of specific clients, is granted by Cold Spring Harbor Laboratory Press, provided that the appropriate fee is paid directly to the Copyright Clearance Center (CCC). Write or call CCC at 222 Rosewood Drive, Danvers, MA 01923 (978-750-8400) for information about fees and regulations. Prior to photocopying items for educational classroom use, contact CCC at the above address. Additional information on CCC can be obtained at CCC Online at www.copyright.com.

For a complete catalog of all Cold Spring Harbor Laboratory Press publications, visit our website at www.cshlpress.org.

Contents

Preface

Genomes encode the rules for life-forms. Differences in genomes underlie most organismal diversity, and aberrations in genomes underlie many disease states. With the rapid advances in DNA sequencing, we now have near-complete genomes for a range of organisms and a fairly comprehensive catalog of human germline and somatic variants, as well as rich annotations of functional genomic elements. The next frontier in the field is to obtain a complete functional annotation of genetic variants and genomic elements at the cellular and organismal levels. Such an understanding, especially in the human context, will not only pave the way for a deeper understanding of the genomic code but will also power therapeutic interventions directed at both effecting cures and eventually also engineering disease resistance. Consequently as we move from reading genomes to interpreting genomes and ultimately engineering genomes, technologies to directly and precisely perturb genomic elements and combinations thereof will be a most critical toolset in these basic science cum engineering endeavors.

In this regard, the recent advent of RNA-guided effectors derived from clustered regularly interspaced short palindromic repeats (CRISPR)–CRISPR-associated systems (Cas) has dramatically transformed our ability to engineer the genomes of diverse organisms. As unique factors capable of colocalizing RNA, DNA, and protein, tools and techniques based on CRISPR–Cas are paving the way for unprecedented control over cellular organization, regulation, and behavior.

Notably, CRISPR–Cas systems evolved as adaptive immune defenses of bacteria and archaea and use short RNA to direct degradation of foreign nucleic acids. They provide immunity by incorporating fragments of invading phage and plasmid DNA into CRISPR loci and using the corresponding CRISPR RNAs (crRNAs) to guide the degradation of homologous sequences. Each CRISPR locus encodes acquired "spacers" that are typically separated by repeat sequences. Transcription of the locus yields a pre-crRNA, which is processed to yield crRNAs that guide effector nuclease complexes to disrupt sequences complementary to the spacer. CRISPR systems are thus readily retargeted by expressing or delivering appropriate crRNAs, and progressive mechanistic insights into these fundamental processes thus paved the way for their recent engineering into a range of prokaryotic and eukaryotic organisms.

In considering the developments in this rapidly evolving field and its applications for understanding basic biology and engineering of new therapeutic paradigms, our goal in developing this book was to highlight the major advances that have been made that have led to the current state of research, while also providing a guide for implementation of these approaches. As such, the book is divided into multiple parts and, focusing specifically on the CRISPR–Cas9 targeting methodology, it details protocols for applications in a range of species and in ex vivo cum in vivo genome targeting scenarios. We begin with an overview of CRISPR–Cas9 biology, followed by computational and experimental protocols for prediction and validation of native and engineered Cas9 orthologs and guide sequences. Toward harnessing the massively multiplexable and scalable genome engineering enabled by this platform, we next detail protocols for constructing CRISPR libraries for effecting large-scale genetic screens in human cell lines. Given the impending applications of CRISPR–Cas in engineering therapeutics, protocols on establishing an adeno-associated virus–based delivery system into cells and mice are provided next. High-resolution assaying of genomic changes induced by this platform are critical for effectively implementing this approach, and thus we also detail highly sensitive polymerase chain reaction (PCR)-based assays to quantify genome-editing events. We follow this with a collection of protocols for precision genome engineering in a range of organisms including yeast, fruit flies, zebrafish, and mice, as well as human induced pluripotent stem cells. We conclude by detailing protocols to enable targeted genome regulation using the CRISPR–Cas9 platform. These chapters provide a comprehensive, in-depth overview of the experimental procedures prevalent in the field. Looking forward, we anticipate the versatility and ease of use afforded by CRISPR–Cas effectors, coupled with their singular ability to bring together RNA, DNA, and protein in a fully programmable fashion, to form the basis of a progressively expanding experimental toolset for the perturbation, regulation, and monitoring of complex biological systems.

We would like to thank the many scientists who have contributed to this book. We are very grateful for their enthusiasm, hard work, and attention to detail in preparing this book, which can serve as a broad resource for technicians, graduate students, postdocs, and any investigator engaged in genetic studies. Special thanks also go to Maryliz Dickerson at Cold Spring Harbor Laboratory Press for helping make this book a reality.

Jennifer Doudna
Prashant Mali

General Safety and Hazardous Material Information

This manual should be used by laboratory personnel with experience in laboratory and chemical safety or students under the supervision of such trained personnel. The procedures, chemicals, and equipment referenced in this manual are hazardous and can cause serious injury unless performed, handled, and used with care and in a manner consistent with safe laboratory practices. Students and researchers using the procedures in this manual do so at their own risk. It is essential for your safety that you consult the appropriate Material Safety Data Sheets, the manufacturers' manuals accompanying equipment, and your institution's Environmental Health and Safety Office, as well as the General Safety and Disposal Cautions in the Appendix for proper handling of hazardous materials in this manual. Cold Spring Harbor Laboratory makes no representations or warranties with respect to the material set forth in this manual and has no liability in connection with the use of these materials.

All registered trademarks, trade names, and brand names mentioned in this book are the property of the respective owners. Readers should please consult individual manufacturers and other resources for current and specific product information.

Appropriate sources for obtaining safety information and general guidelines for laboratory safety are provided in the General Safety and Hazardous Material Information Appendix.

CHAPTER 1

Overview of CRISPR–Cas9 Biology

Hannah K. Ratner,[1,2,3,6] Timothy R. Sampson,[1,2,3,5,6] and David S. Weiss[2,3,4,7]

[1]*Department of Microbiology and Immunology, Microbiology and Molecular Genetics Program, Emory University, Atlanta, Georgia 30329;* [2]*Emory Vaccine Center, Emory University, Atlanta, Georgia 30329;* [3]*Yerkes National Primate Research Center, Emory University, Atlanta, Georgia 30329;* [4]*Division of Infectious Diseases, Department of Medicine, Emory University School of Medicine, Atlanta, Georgia 30329*

Prokaryotes use diverse strategies to improve fitness in the face of different environmental threats and stresses, including those posed by mobile genetic elements (e.g., bacteriophages and plasmids). To defend against these elements, many bacteria and archaea use elegant, RNA-directed, nucleic acid–targeting adaptive restriction machineries called CRISPR–Cas (CRISPR-associated) systems. While providing an effective defense against foreign genetic elements, these systems have also been observed to play critical roles in regulating bacterial physiology during environmental stress. Increasingly, CRISPR–Cas systems, in particular the Type II systems containing the Cas9 endonuclease, have been exploited for their ability to bind desired nucleic acid sequences, as well as direct sequence-specific cleavage of their targets. Cas9-mediated genome engineering is transcending biological research as a versatile and portable platform for manipulating genetic content in myriad systems. Here, we present a systematic overview of CRISPR–Cas history and biology, highlighting the revolutionary tools derived from these systems, which greatly expand the molecular biologists' toolkit.

INTRODUCTION AND HISTORY

For decades, the function and purpose of CRISPR (clustered, regularly interspaced, short, palindromic repeats)–Cas (CRISPR-associated) systems remained an enigma, until a series of astute observations paved the way for an exploding field of research on the biology of these prokaryotic adaptive immune systems and the exploration of how they can be exploited for directed genome modification. The rapid evolution of this field has been dubbed the "CRISPR craze" and is widely recognized throughout the scientific community as having already revolutionized genetic engineering (Pennisi 2013; Barrangou 2014; Doudna and Charpentier 2014). Only 3 years after the first proof-of-principle experiments demonstrating that these systems could be reprogrammed and exploited as genome engineering tools, Cas9 technologies have not only been used to generate genetic knockout mutants in diverse organisms and model systems, but for a variety of other applications including, but not limited to, transcriptional repression and activation and live-cell imaging of DNA localization (Jinek et al. 2012; Chen et al. 2013; Cong et al. 2013; DiCarlo et al. 2013; Jiang et al. 2013; Mali et al. 2013; Perez-Pinera et al. 2013; Qi et al. 2013; Doudna and Charpentier 2014; Sampson and Weiss 2014).

It is a little-known fact that the study of CRISPR–Cas systems unknowingly began more than 25 years ago when an array of short, repetitive DNA sequences (~20–40 bp in length, termed "repeats") inter-

[5]Present address: Division of Biology and Biological Engineering, California Institute of Technology, Pasadena, California 91125

[6]These authors contributed equally to this work.

[7]Correspondence: david.weiss@emory.edu

Copyright © Cold Spring Harbor Laboratory Press; all rights reserved

Cite this introduction as *Cold Spring Harb Protoc;* doi:10.1101/pdb.top088849

spaced with nonrepetitive sequences (termed "spacers") was identified following the sequencing of the gene encoding alkaline phosphatase isozyme conversion enzyme (*iap*) in the *Escherichia coli* genome (Ishino et al. 1987). At the time, the function and purpose of these sequences were unknown. However, two decades later, computational analyses led to the discovery that these repetitive arrays were present in numerous bacteria and archaea and, notably, that the spacers were identical to many sequences present in exogenous mobile genetic elements such as plasmids, transposons, and bacteriophages (Bolotin et al. 2005; Mojica et al. 2005). Further bioinformatic studies revealed that these arrays, termed CRISPR arrays, were often associated with a core set of Cas genes (Jansen et al. 2002; Haft et al. 2005). Many of the Cas genes had sequence similarity to endonuclease and helicase families or genes encoding other nucleic acid binding proteins (Jansen et al. 2002; Haft et al. 2005; Makarova et al. 2006). In conjunction with the fact that many spacers were identical to mobile genetic elements, these findings gave rise to the postulation that CRISPR–Cas systems may act as a form of RNA-directed interference against foreign genetic elements (Makarova et al. 2006). This hypothesis was solidified in 2007 by a set of foundational experiments that provided the first direct evidence that CRISPR sequences and the associated Cas proteins directed interference against bacteriophage infection (Barrangou et al. 2007). Perhaps even more interestingly, new spacer sequences were naturally acquired into the CRISPR array following bacteriophage infection, subsequently facilitating sequence-specific resistance to the offending phage, and revealing a mechanism of adaptive immunity in prokaryotes (Barrangou et al. 2007; Brouns et al. 2008; Gasiunas et al. 2012; Westra et al. 2012; Staals et al. 2013).

Over the last 8 years, the mechanism of RNA-directed interference by CRISPR–Cas systems has been largely uncovered (Barrangou and Marraffini 2014; Plagens et al. 2015; Rath et al. 2015). Briefly, CRISPR-mediated interference occurs in three primary stages: (1) spacer acquisition, (2) crRNA transcription and maturation, and (3) target identification and cleavage (Fig. 1). During spacer acquisition, foreign nucleic acids are identified and processed into short, spacer-sized sequences that are inserted into the CRISPR array, to be flanked by a pair of repeat sequences (Fig. 1A–D; Heler et al. 2014). The CRISPR array is then transcribed and processed into mature small RNAs, called crRNAs, that each contain portions of the repeat sequences and a single spacer that facilitates identification of a target nucleic acid with significant sequence complementarity to the spacer sequence (Fig. 1E,F). The crRNAs complex with Cas protein(s) and, in some cases, additional RNAs to bind the target, resulting in target cleavage (Fig. 1G,H; Barrangou and Marraffini 2014; Plagens et al. 2015; Rath et al. 2015).

The field of CRISPR–Cas biology continues to rapidly expand. Numerous groups have elegantly revealed not only the molecular function of CRISPR–Cas systems in defense against foreign nucleic acids (Barrangou et al. 2007; Brouns et al. 2008; Marraffini and Sontheimer 2008; Hale et al. 2009; Garneau et al. 2010; Bikard et al. 2012; Gasiunas et al. 2012) but also uncovered clues about the evolution of these systems (Makarova et al. 2011; Chylinski et al. 2014; Krupovic et al. 2014; Koonin and Krupovic 2015) and their functions in other physiological processes (Bikard and Marraffini 2013; Westra et al. 2014; Barrangou 2015; Ratner et al. 2015). Most recently, and as is the topic of this collection, this foundational work has led to the discovery of how these systems, and specifically the CRISPR-associated endonuclease Cas9, can be engineered for myriad biotechnological applications.

TYPES OF CRISPR–Cas SYSTEMS

CRISPR–Cas systems can be subdivided into three main types (Type I, II, and III) that are each distinguished by the presence of unique Cas proteins, encoded adjacent to the CRISPR array (Makarova et al. 2011). Despite their conserved function in prokaryotic adaptive immunity, CRISPR–Cas systems are structurally and mechanistically diverse (Makarova et al. 2011, 2013; Vestergaard et al. 2014). The adaptation stage of immunity is the most conserved between the three CRISPR–Cas subtypes, with all known systems encoding the Cas proteins involved in this process, Cas1 and Cas2 (Fig. 1A–C; Heler et al. 2014). These two metal-dependent nucleases are both necessary and sufficient for spacer acquisition, but dispensable for target interference (Datsenko et al. 2012; Yosef et al. 2012; Nunez et al. 2014, 2015; Heler et al. 2015). Recently solved crystal structures of Cas1 and

 Cite this introduction as *Cold Spring Harb Protoc*; doi:10.1101/pdb.top088849

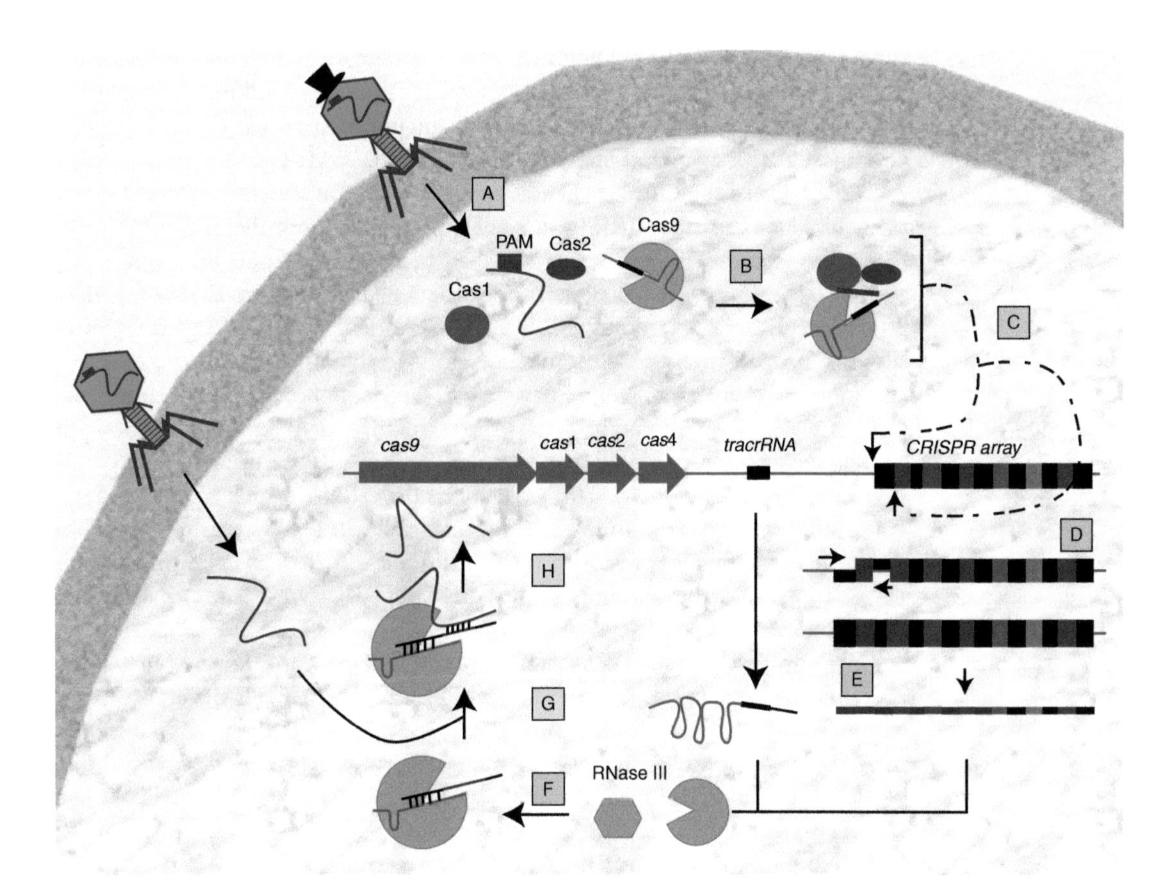

FIGURE 1. The three stages of adaptive immunity by Type II-C CRISPR–Cas systems. (*A–D*) Spacer acquisition: (*A*) foreign DNA (dark purple) enters the cell, and (*B*) *Cas1*, *Cas2*, and *Cas9* in complex with tracrRNA (blue) select a spacer sequence on the target through Cas9-mediated identification of a protospacer adjacent motif (PAM; dark purple rectangle on the foreign DNA). The PAM adjacent sequence is processed into a spacer-sized fragment. (*C*) The Cas protein complex attached to the spacer identifies the CRISPR array and creates staggered single-stranded breaks on each side on a repeat. (*D*) The new spacer sequence is inserted into the array and the single-stranded repeats on either side of the new spacer are repaired by DNA polymerase I. (*E,F*) crRNA transcription and maturation: (*E*) the CRISPR array and tracrRNA are transcribed. (*F*) Cas9 binds tracrRNA and the CRISPR transcript, which is then cleaved into mature, spacer-specific crRNAs by RNase III. The mature dual crRNA:tracrRNA remains bound to Cas9 as a heteroduplex. (*G,H*) Target identification and cleavage: (*G*) Upon re-infection with foreign DNA, the spacer on the crRNA of the Cas9:RNA heteroduplex binds to its complementary sequence on the foreign nucleic acid. (*H*) Cas9 adopts a conformationally active state and cleaves both DNA strands in the target, protecting the cell.

Cas2 indicate that these proteins form stable, heterodimeric complexes in vitro, and that in vivo, the interaction between Cas1 and Cas2 is necessary for recognizing the DNA secondary structure of the CRISPR repeat sequence during integration of new spacers (Nunez et al. 2014). The catalytic activity of Cas1 is essential for spacer acquisition, whereas the predicted nuclease active site of Cas2 is not (Nunez et al. 2014, 2015). Evidence from multiple types of CRISPR–Cas systems indicates that Cas1 and Cas2 may form complexes with Cas proteins involved in target identification and cleavage (Datsenko et al. 2012; Plagens et al. 2012; Swarts et al. 2012; van der Oost et al. 2014; Heler et al. 2015; Wei et al. 2015). Spacer acquisition may require these other Cas proteins to accurately select sequences in a way that prevents the CRISPR–Cas system from targeting its own chromosomal spacer sequences with the crRNAs transcribed from it; the details of this are described in sections below for the Type II systems (Barrangou et al. 2007; Datsenko et al. 2012; Heler et al. 2015).

The differences between the distinct types of CRISPR–Cas systems become increasingly clear at the crRNA maturation, target identification, and interference stages of immunity. Notably, Type I and III systems (described in this section) use large, multimeric protein complexes for these activities, whereas the Type II systems (described in detail in subsequent sections) require a single protein for these diverse functions (van der Oost et al. 2014). Type I systems use the endonucleases Cas6 or Cas5d to cleave the CRISPR array transcript within the repeat sequences flanking each spacer, resulting in a short 5′ repeat-derived sequence and a 3′ hairpin, including a repeat-derived sequence (Carte et al. 2010; Gesner et al. 2011; Jore et al. 2011; Sashital et al. 2011; Garside et al. 2012; Nam et al. 2012; Koo et al. 2013; Reeks et al. 2013). The Cas6 protein then transports the mature crRNA to a complex of Cas proteins called Cascade (CRISPR-associated complex for antiviral defense), which functions in interference, in some cases remaining attached to the crRNA and becoming a part of the interference complex (Brouns et al. 2008; Carte et al. 2008; Haurwitz et al. 2010, 2012; Hatoum-Aslan et al. 2011; Jore et al. 2011; Wang et al. 2011; Sternberg et al. 2012; Niewoehner et al. 2014). Type I systems form an interference complex of four to five distinct Cas proteins, each with different stoichiometry (Brouns et al. 2008; Jore et al. 2011; Sashital et al. 2011; Nam et al. 2012; van Duijn et al. 2012). Cryoelectron microscopy (CryoEM) structures of this complex indicate that six copies of Cas7, a protein with a ferredoxin fold that resembles an RNA Recognition Motif, form an RNA-binding ridge (Wiedenheft et al. 2011a, b). This ridge binds the crRNA, which is anchored by other Cas proteins on either end of the Cas7 multimer (Lintner et al. 2011). When the crRNA binds the target DNA, conformational changes result in the recruitment of the Cas3 endonuclease, which mediates target degradation and is the defining Cas protein of Type I systems (Jore et al. 2011; Wiedenheft et al. 2011b; Westra et al. 2012).

Like the Type I systems, Type III systems also use Cas6 for crRNA processing and form multiprotein complexes for target interference (Reeks et al. 2013). However, the Cas proteins in the Type III complexes are different (Spilman et al. 2013; Staals et al. 2013). Cas10 is a component of Type III interference complexes and is the defining Cas protein of these systems, although its function has not been fully elucidated (Makarova et al. 2011). CryoEM structures of Type III systems show that the crRNA is positioned along a backbone of a Cas protein complex consisting of repeat units of Csm3 (III-A) or Cmr4 (III-B), much like the Cas7 repeats in Type I systems (Zhang et al. 2012; Hrle et al. 2013; Rouillon et al. 2013; Spilman et al. 2013; Staals et al. 2013, 2014). Interestingly, both Type III-A and III-B systems are capable of targeting DNA and RNA (Hale et al. 2009; Peng et al. 2015; Samai et al. 2015). In Type III-A systems, degradation of DNA requires Cas10 and cleavage occurs directly adjacent to the 3′ end of the bound crRNA (Samai et al. 2015). Degradation of RNA targets by Type III-A systems occurs in even, 6-nucleotide intervals via the Csm3 active site, with each identical subunit in the backbone individually cleaving the target to collectively fragment the invading nucleic acid into consistent and precisely sized sequences (Staals et al. 2013; Samai et al. 2015). It is likely that the backbone repeat of Cmr4 in Type III-B systems has a similar mechanism of target cleavage (Staals et al. 2013; van der Oost et al. 2014).

Specificity of the crRNA for the target is enhanced through distinct mechanisms in different systems to avoid off-target effects that could occur because of binding of fully or partially complementary sequences, as mistargeting of the host chromosome is likely lethal to the bacteria. Types I and II systems improve specificity through recognition of a specific nucleotide sequence adjacent to the target but on the complementary strand of DNA, called the PAM (protospacer adjacent motif) (Bolotin et al. 2005; Deveau et al. 2008; Mojica et al. 2009; Marraffini and Sontheimer 2010). PAM recognition facilitates Cas interference complex binding, DNA melting, and RNA:DNA heteroduplex formation (described in detail below for Type II systems) and prevents self-targeting of similar or identical sequences lacking a PAM (Marraffini and Sontheimer 2010). Interestingly, some Type III-A systems may avoid cleavage of sequences incorporated into the host genome through a unique transcription-dependent DNA targeting mechanism that enables tolerance of lysogenic phages while preventing lytic phage production (Goldberg et al. 2014).

Cas9-MEDIATED crRNA MATURATION

In contrast to Type I and III systems, Type II systems require a single Cas protein, the Cas9 endonuclease, to mediate crRNA maturation (Deltcheva et al. 2011). The CRISPR array is first transcribed as a single, long transcript. Subsequently, this pre-crRNA transcript is processed into individual crRNAs, each specific for a different target (Fig. 1E,F). A single, matured, spacer-specific crRNA is then complexed with Cas9 as well as the *trans*-activating crRNA (tracrRNA), a small RNA encoded within the CRISPR–Cas locus, and unique to Type II systems. The tracrRNA contains multiple stem-loop structures and a sequence with partial complementarity to the CRISPR repeat sequence, allowing binding to the crRNA to facilitate maturation and complex formation with Cas9 (Deltcheva et al. 2011; Jinek et al. 2012; Chylinski et al. 2013, 2014; Fonfara et al. 2014). The dsRNA endonuclease, RNase III, which is typically encoded distal from the CRISPR locus, is also required for crRNA maturation (Deltcheva et al. 2011). RNase III recognizes the dsRNA structure created by the tracrRNA:crRNA duplex and cleaves both strands of RNA within the double-stranded repeat region (Deltcheva et al. 2011). The tracrRNA:crRNA duplex binds tightly to Cas9 and undergoes additional processing through an unknown mechanism that likely involves additional bacterial RNases (Deltcheva et al. 2011). The dual RNA:Cas9 complex is then able to identify and cleave targets with sequence complementarity to the crRNA spacer (Fig. 1G,H; Deltcheva et al. 2011; Gasiunas et al. 2012; Jinek et al. 2012; Chylinski et al. 2013; Fonfara et al. 2014). In some Type II systems, notably that encoded by the pathogen *Neisseria meningitidis*, maturation of the crRNAs is independent of RNase III and tracrRNA (Zhang et al. 2013). In this case, internal promoter sequences within each repeat sequence allow for transcription of individual crRNAs. These crRNAs still require tracrRNA to associate with Cas9, highlighting the importance of the RNA duplex for interactions with this protein (Zhang et al. 2013).

TARGET INTERFERENCE BY Cas9

The mechanism of target interference by Type II CRISPR–Cas systems has been well established and sophisticatedly elucidated, greatly informed by the solving of the crystal structures of Cas9 alone and bound to DNA and RNA (Deltcheva et al. 2011; Gasiunas et al. 2012; Jinek et al. 2012, 2014; Fonfara et al. 2014; Nishimasu et al. 2014). Similar to its role in crRNA maturation, Cas9 is the sole Type II Cas protein involved in target surveillance and interference (Deltcheva et al. 2011; Jinek et al. 2012).

Cas9 has a two-lobed morphology, with a larger α-helical lobe and smaller nuclease lobe that together form a clam-like shape with a central channel to position the target (Fig. 2A,B; Jinek et al. 2014; Nishimasu et al. 2014). Cas9 first binds the crRNA:tracrRNA duplex via a positively charged arginine-rich motif located on the inner surface of the α-helical lobe, where the two lobes come together at the end of the central cavity (Jinek et al. 2014; Nishimasu et al. 2014). Upon RNA binding, Cas9 undergoes a first conformational change to create the channel that positions the nucleic acids along the length of the protein, by rotating the nuclease lobe around the nucleic acid binding pocket of the α-helical lobe (Jinek et al. 2014; Nishimasu et al. 2014). This reorients the endonuclease domains to either side of the channel, into a favorable conformation for subsequent target cleavage (Figs. 1G,H and 2B,C) (Jinek et al. 2014; Nishimasu et al. 2014).

Cas9 must then scan DNA to identify target sequences with a high degree of accuracy so as not to target its own chromosome. This is partially accomplished by the requirement for the PAM motif (typically ~3 bp) adjacent to the targeted region on the target DNA (Figs. 1 and 2B,C) (Gasiunas et al. 2012; Jinek et al. 2012; Fonfara et al. 2014). Cas9 associates and dissociates randomly along a DNA strand until encountering a PAM sequence (Sternberg et al. 2014). Subsequently, the PAM-interacting domain of Cas9 (located in the carboxyl terminus) binds tightly to the target DNA through two binding loops that interact with the major and minor grooves of the PAM (Jinek et al. 2014; Nishimasu et al. 2014). Cas9 then undergoes a second conformational change, locking the DNA target into place along the length of the central cavity between the two lobes (Jinek et al. 2014; Nishimasu et al. 2014). Interaction with the PAM leads to destabilization of adjacent double-stranded DNA and orients the target sequence to facilitate binding to the seed region of the crRNA (Jinek et al. 2014; Nishimasu et al. 2014). If the target sequence has near-perfect complementarity in the PAM-proximal

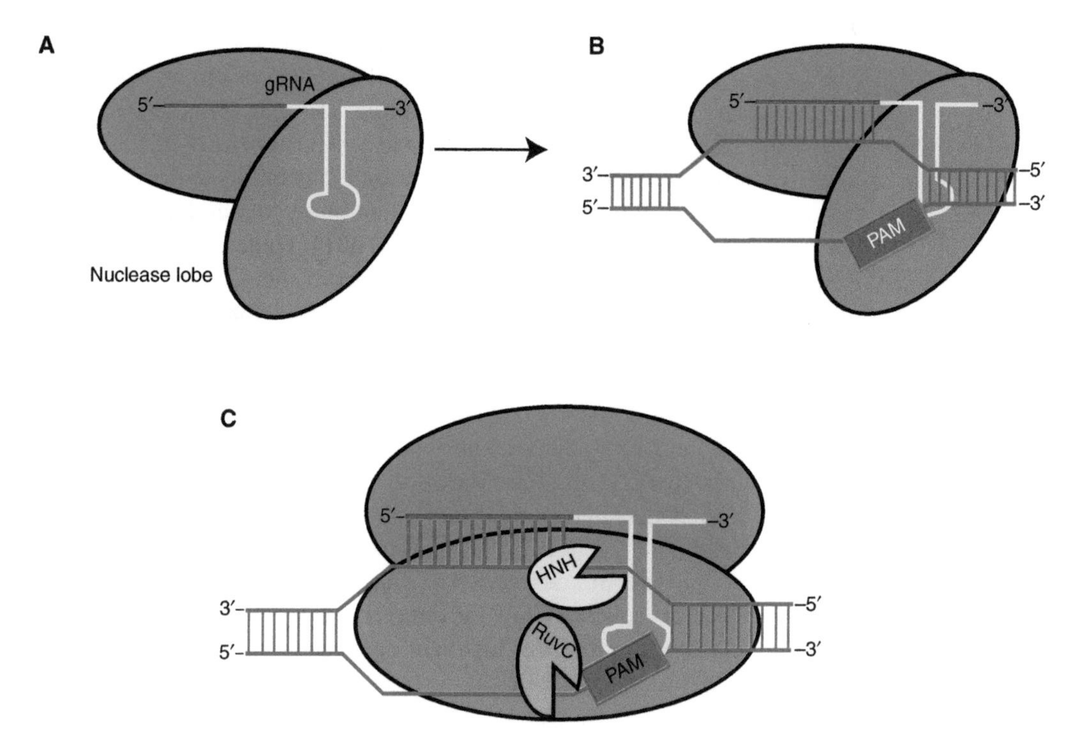

FIGURE 2. Schematic of Cas9:gRNA interactions. (*A*) Upon association with a chimeric gRNA, consisting of an ssRNA targeting region similar to the crRNA (red) and a dsRNA structure similar to that created by the crRNA:tracrRNA complex (yellow), the α-helical lobe (blue) and the nuclease lobe (pink) of Cas9 are opened into a conformation that reveals a channel for DNA targets to bind. (*B*) When DNA containing a PAM sequence is identified by Cas9, and the targeting sequence of the gRNA (red) has significant sequence complementarity to the immediately adjacent DNA sequence, the DNA is melted and unwound, generating a DNA:RNA hybrid. (*C*) Cas9 then undergoes a conformational change, clamping its nuclease lobe across the targeted DNA and positioning each strand into the HNH and RuvC active sites of the nuclease lobe. The HNH and RuvC endonuclease domains then cleave the complementary and noncomplementary strands, respectively, resulting in a double-strand break in the target immediately adjacent to the PAM.

region of the spacer, melting along the DNA will occur as one strand of the target base pairs along the remainder of the complementary spacer, forming an RNA:DNA heteroduplex (Anders et al. 2014; Jinek et al. 2014; Nishimasu et al. 2014). This results in separation of the two DNA strands into distinct, metal ion-dependent endonuclease active sites (Jinek et al. 2014; Nishimasu et al. 2014).

The HNH endonuclease domain cleaves the DNA strand bound to the RNA three nucleotides upstream of the PAM, whereas the noncomplementary strand is also bound by the nuclease lobe of Cas9 but cleaved by a separate RuvC domain (Jinek et al. 2012, 2014; Nishimasu et al. 2014). These active sites preferentially use magnesium as a divalent ion but can tolerate manganese (although with a lower cleavage efficiency), whereas calcium inhibits activity (Jinek et al. 2012; Anders et al. 2014). Interestingly, recent in vitro kinetic studies suggest that Cas9 is a single turnover enzyme that remains bound to the DNA target following cleavage, and the fate of Cas9 that has completed cleavage is currently unknown (Sternberg et al. 2014).

SPACER ACQUISITION IN Cas9-DEPENDENT CRISPR–Cas SYSTEMS

Adaptation, through the acquisition of new spacers into the CRISPR array, is the least understood stage of canonical CRISPR–Cas function. In Type II-A systems, all components of the CRISPR–Cas system form a complex that is required for adaptation (Cas1, Cas2, Cas9, Csn2, and tracrRNA) (Heler et al. 2015; Wei et al. 2015). A similar mechanism is likely used by other Type II subtypes that contain these components, excluding Csn2, which is absent from Type II-C and is replaced by Cas4 in Type II-

Cite this introduction as *Cold Spring Harb Protoc*; doi:10.1101/pdb.top088849

B subtypes (Chylinski et al. 2013, 2014). Both Csn2 and Cas4 resemble RecB-like nucleases and may therefore play a similar role in adaptation, although their precise functions are not known (van der Oost et al. 2014). Csn2 and Cas4, as well as Cas1 and Cas2, are all dispensable for crRNA processing and target interference in Type II CRISPR–Cas systems (Deltcheva et al. 2011; Jinek et al. 2012). Interestingly, the Cas1 proteins present in Type II CRISPR–Cas systems cluster phylogenetically with those of Type I systems (Chylinski et al. 2014). This may indicate that the distinct functions of Type II systems arose via recombination events with Cas9 and other types of CRISPR–Cas systems, such as the Type I system (Chylinski et al. 2014).

Upon invasion by a foreign nucleic acid, CRISPR–Cas systems must select spacer sequences in a manner that prevents autoimmunity (Stern et al. 2010; Heler et al. 2015). Type II systems accomplish this by requiring a specific PAM sequence adjacent to the one that will ultimately be integrated as the spacer (i.e., the protospacer) (Díez-Villaseñor et al. 2013; Nunez et al. 2014; Heler et al. 2015). In Type II-A systems, Cas9, in complex with Cas1, Cas2, and Csn1 and bound to tracrRNA, identifies PAMs on the invading DNA to facilitate spacer selection using the PAM-interacting domain (Jinek et al. 2014; Nishimasu et al. 2014; Heler et al. 2015; Wei et al. 2015). There may be additional requirements for the selection of the spacer sequence, as there is an enrichment for certain spacer sequences that cannot be accounted for by the sequence of the PAM alone; however, these requirements have yet to be identified (Heler et al. 2014).

Mutations in the PAM-interacting domain of Cas9 do not prevent spacer acquisition but instead result in incorporation of spacers that are not adjacent to a PAM in the target (Heler et al. 2015). The endonuclease activity of Cas9 is dispensable for acquisition, suggesting that the role for Cas9 is to select spacers by binding to the PAM and protospacer sequence, whereas Cas1 (whose nonspecific nuclease activity is required for adaptation) of the associated Cas1–Cas2–Csn1 complex cleaves the adjacent sequence, yielding a precisely selected spacer sequence (Heler et al. 2015). There are many unknowns in the mechanism of adaptation, but a general model has been developed (Fig. 1A–D; Heler et al. 2014, 2015; Nunez et al. 2014, 2015). Cas1–Cas2 together interact with the secondary structures of the CRISPR repeat sequences within the array, preferentially near the leader sequence, which also acts as a promoter (Nunez et al. 2014, 2015). A repeat sequence within the chromosomal array is then nicked at the 3′ end, allowing for ligation of the free hydroxyl to the spacer fragment (Nunez et al. 2015). The spacer is inserted into the array, flanked by the single complementary strands of the first CRISPR repeat (Nunez et al. 2015). These are repaired into double-stranded repeats by DNA polymerase, resulting in a new repeat-flanked spacer in the chromosome, to be transcribed and processed into a crRNA that can protect against future invasion by complementary, PAM-flanked sequences (Nunez et al. 2015).

ALTERNATIVE FUNCTIONS OF Cas9 IN BACTERIAL PHYSIOLOGY

Although CRISPR–Cas systems have been very well established to promote prokaryotic defense against foreign nucleic acids, there is increasing evidence that these systems, and Cas9 in particular, play important roles in bacterial physiology (Bikard and Marraffini 2013; Westra et al. 2014; Barrangou 2015; Ratner et al. 2015). These additional Cas9-mediated functions include endogenous gene regulation and facilitate the strengthening of envelope structure, resistance to antibiotics and ultimately allow certain bacterial pathogens to dampen host immune activation (Bikard and Marraffini 2013; Westra et al. 2014; Barrangou 2015; Ratner et al. 2015).

Some alternative Cas9 functions have been revealed through the study of the intracellular pathogen *Francisella novicida* (Sampson and Weiss 2013). Using a regulatory axis comprised of Cas9, tracrRNA, and a unique small RNA encoded adjacent to the CRISPR array, the scaRNA (small, CRISPR–Cas associated RNA), *F. novicida* represses the production of a specific endogenous bacterial lipoprotein (BLP) (Sampson et al. 2013; Chylinski et al. 2014). Repression of this BLP by the *F. novicida* Cas9 regulatory axis allows the bacterial cell to strengthen the integrity of its envelope, decreasing envelope permeability and promoting resistance to certain antibiotics (Sampson et al.

2014). Furthermore, because BLPs are recognized by the host innate immune receptor, Toll-like receptor 2 (TLR2), repression of BLP allows *F. novicida* to dampen the activation of TLR2 and prevent inflammatory immune signaling, ultimately promoting survival and replication in the host (Jones et al. 2012; Sampson et al. 2013, 2014). The precise mechanism of Cas9-mediated gene repression in this system is unknown. In the absence of Cas9, tracrRNA, or the scaRNA, levels of the BLP transcript are drastically increased, as these components together act to decrease the stability of the mRNA (Sampson et al. 2013). Interestingly, the catalytic residues within Cas9 that are involved in DNA cleavage are not essential to maintain low levels of BLP transcript, most likely suggesting that stability is altered by currently unidentified accessory RNases (Heidrich and Vogel 2013; Sampson et al. 2013).

F. novicida is not the only bacterium known to use Cas9 in a fashion distinct from defense against invading nucleic acid. Cas9 encoded by *Neisseria meningitidis* is necessary for attachment, entry, and intracellular survival of the bacteria in human epithelial cells (Sampson et al. 2013). *Campylobacter jejuni* also uses Cas9 for attachment and invasion of epithelial cells (Louwen et al. 2013). In the absence of Cas9, *C. jejuni* displays increased envelope permeability, antibiotic susceptibility, and surface antibody binding, which may suggest that Cas9 acts to regulate components of the *C. jejuni* envelope (Louwen et al. 2013; Sampson et al. 2014). Numerous other examples of alternative CRISPR–Cas activities that do not use Cas9 have been observed and are growing in number. Some notable examples include biofilm formation in *Pseudomonas aeruginosa* (Zegans et al. 2009; Cady and O'Toole 2011), fruiting body formation in *Myxococcus xanthus* (Viswanathan et al. 2007; Wallace et al. 2014), intra-amoeba survival in *Legionella pneumophila* (Gunderson and Cianciotto 2013; Gunderson et al. 2015), colonization of nematodes by *Xenorhabdus nematophila* (Veesenmeyer et al. 2014), and many others. However, as these systems do not use Cas9, they are beyond the scope of this discussion and have been reviewed extensively elsewhere (Bikard and Marraffini 2013; Westra et al. 2014; Barrangou 2015; Ratner et al. 2015). Nonetheless, such examples of moonlighting functions of CRISPR–Cas systems in bacterial physiology may provide the framework to understand the evolution of these systems as well as how they may be further used and exploited for biotechnological purposes.

USE OF Cas9 FOR GENOME ENGINEERING

The insights into the mechanism of Cas9 function led to the hypothesis that the spacer sequence of the crRNA targeting region could be reprogrammed such that this machinery would mediate target cleavage at sites of interest. This activity was subsequently shown, in vitro, only 5 years after the first functional description of these systems (Barrangou et al. 2007; Jinek et al. 2012). It was clearly shown that synthetic crRNAs could be produced that were capable of hybridizing to DNA sequences of interest, allowing *Streptococcus pyogenes* Cas9 to catalyze a double-stranded DNA break at that site (Jinek et al. 2012). Although still exceedingly less complex than synthetically engineering other site-specific nucleases, such as zinc finger nucleases or TALENs (transcription-like effector nucleases), the expression of two separate small RNAs nonetheless represented added difficulty. The requirement for tracrRNA was relieved and the system simplified even further with the engineering of a synthetic, double-stranded targeting RNA (a guide RNA, or gRNA) (Jinek et al. 2012). (See Chapter 2 Introduction: Guide RNAs: A Glimpse at the Sequences that Drive CRISPR–Cas Systems [Briner and Barrangou 2016].) The gRNA retains the double-stranded sequence and structural elements of the tracrRNA:crRNA duplex that are necessary for interaction with Cas9 but is transcribed as a single RNA (Jinek et al. 2012; Chylinski et al. 2013). This chimeric RNA therefore does not require RNase III processing. The spacer sequence, which directs Cas9 targeting, can easily be modified, facilitating reprogramming against diverse targets (Deltcheva et al. 2011; Jinek et al. 2012). The generation of the gRNA significantly increased the ease of engineering new targeting sequences and, together with the elucidation of Cas9 activity, helped pave the way for a revolutionary, highly cost-effective, and efficient method of genome engineering.

 Cite this introduction as *Cold Spring Harb Protoc*; doi:10.1101/pdb.top088849

These developments have now sprung the so-called CRISPR craze of Cas9-mediated genome engineering in many systems, both prokaryotic and eukaryotic. Cas9 from multiple bacterial species (including *S. pyogenes*, *S. thermophilus*, *Staphylococcus aureus*, *N. meningitidis*, and *Treponema denticola*) have been successfully used to edit the genomes of cells from diverse organisms including the human (discussed in Chapter 11 Protocol 1: A Method for Genome Editing in Human Pluripotent Stem Cells [Smith et al. 2016]), bacteria, yeast (discussed in Chapter 7 Protocol 1: CRISPR–Cas9 Genome Engineering in *Saccharomyces cerevisiae* Cells [Ryan et al. 2016]), nematode, plants, fruitfly (discussed in Chapter 8 Introduction: Cas9-Mediated Genome Engineering in *Drosophila melanogaster* [Housden and Perrimon 2016]), zebrafish (discussed in Chapter 9 Protocol 1: Optimized CRISPR–Cas9 System for Genome Editing in Zebrafish [Vejnar et al. 2016]), salamander, frog, and rodent (discussed in Chapter 10 Protocol 1: Generation of Genetically Modified Mice Using the CRISPR–Cas9 Genome-Editing System [Henao-Mejia et al. 2016]), with target modification efficiencies reported up to 80% (Jinek et al. 2012, 2013; Belhaj et al. 2013; Cho et al. 2013; Cong et al. 2013; DiCarlo et al. 2013; Gratz et al. 2013; Hou et al. 2013; Hwang et al. 2013; Jiang et al. 2013; Lo et al. 2013; Nakayama et al. 2013; Nekrasov et al. 2013; Ren et al. 2013; Wang et al. 2013; Yu et al. 2013; Flowers et al. 2014; Ryan and Cate 2014; Ran et al. 2015). Such rapid utilization across these varied systems serves to highlight the ease of use and portability of Cas9-based technologies.

In the simplest use of Cas9 genome editing, random mutations are introduced at the site of cleavage. Because Cas9 catalyzes a double-strand break at its cleavage site adjacent to the PAM, cells can undergo nonhomologous end joining (NHEJ) to repair the cleaved DNA (Cong et al. 2013; DiCarlo et al. 2013; Jinek et al. 2013). With varying efficacies based on the cellular repair machineries, NHEJ can restore the cleaved sequence to the original, but it can also result in the loss or addition of nucleotides (Cong et al. 2013; DiCarlo et al. 2013; Jinek et al. 2013). The majority of mutations that are generated following Cas9-mediated cleavage are either single-base insertions or deletions or nine-base deletions (Cradick et al. 2013). Such NHEJ-mediated repair can therefore result in early stop codons or other frameshift mutations that can cause loss of function of the targeted gene. Ultimately, this can provide a quick and simple method to generate null mutations in genes of interest.

An alternative repair pathway to NHEJ can also occur within the cell, termed homology-directed repair (HDR). HDR transpires when DNA containing sequence homology with the region surrounding the cleavage site is used as a template for homologous recombination. By introducing linear or circular DNA containing a sequence of interest (such as a selectable or nonselectable marker) flanked by regions homologous to those adjacent to the Cas9 cleavage site, integration of this donor construct can occur by HDR. This allows Cas9 to effectively generate desired insertions of DNA into sequence-specific sites of interest (Cong et al. 2013; DiCarlo et al. 2013). (The detection of HDR events is discussed in Chapter 6 Introduction: Detecting Single-Nucleotide Substitutions Induced by Genome Editing [Miyaoka et al. 2016].) Furthermore, to increase the likelihood of HDR and limit the chances of NHEJ, a partially mutated Cas9 protein can be used. Engineered point mutations in either one of the two Cas9 endonuclease domains (RuvC or HNH) results in a protein that is capable of only cleaving a single strand of its DNA target (Jinek et al. 2012; Cong et al. 2013; DiCarlo et al. 2013). This decreases the frequency of NHEJ repair, and in the presence of a donor construct, these single-strand nicks are preferentially repaired by HDR. To further increase the rate of HDR, NHEJ can be inhibited (Chu et al. 2015; Maruyama et al. 2015). This has successfully been accomplished by either transcriptionally silencing the NHEJ machinery or through a small molecule inhibitor of the NHEJ polymerase (DNA Pol IV). By blocking NHEJ, HDR repair rates have been increased by four- to 19-fold, facilitating much more efficient integration of desired sequences into targeted sites (Chu et al. 2015; Maruyama et al. 2015).

The ability to easily target Cas9 to diverse sequences within the same cell allows large-scale screens of genetic knockouts to be performed (a process described in Chapter 4 Introduction: Large-Scale Single Guide RNA Library Construction and Use for Genetic Screens [Wang et al. 2016]), a method previously relegated to the world of prokaryotic genetics. Recent studies have used pools of more than 70,000 gRNAs in both positive and negative screens (Bell et al. 2014; Shalem et al. 2014; Wang et al. 2014; Yin et al. 2014; Zhou et al. 2014). Cas9-based screens allow genes to be fully inactivated, not only repressed as occurs during canonical RNA interference-based screens. This loss-of-function method may allow the identification of genes that maintain functional roles even when repressed to very low expression levels through RNAi methods. Cas9 deletion screens, therefore, will potentially uncover previously masked functions of critical genes.

Although the ability of Cas9 to catalyze sequence-specific DNA breaks has revolutionized the introduction of insertions and deletions into DNA, a number of other technologies have been invented that exploit Cas9's ability to bind and strongly associate with desired DNA sequences. Cas9 can be engineered to be completely catalytically inactive through alanine substitutions in both the RuvC and HNH domains, resulting in a variant termed nuclease-deficient Cas9, or dCas9 (Jinek et al. 2012; Jiang et al. 2013; Qi et al. 2013). dCas9 binds targeted DNA sequences as specified by the gRNA, but rather than cleaving the target, instead prevents transcription by blocking the binding or elongation of RNA polymerase (Jiang et al. 2013; Qi et al. 2013); see Chapter 12 Protocol 1: CRISPR Technology for Genome Activation and Repression in Mammalian Cells (Du and Qi 2016). The level of transcriptional inhibition, or CRISPR interference (CRISPRi), can be tuned with different strategies to titrate the expression level of a transcript. Simultaneously targeting dCas9 to multiple sites in the same gene increases repression, as does increasing the proximity of dCas9 binding to the promoter (Jiang et al. 2013; Qi et al. 2013). Whereas repression can occur via dCas9 alone, this protein can be tethered to other proteins and molecules to facilitate increased efficacy or perform other actions at discrete sites in a genome. Fusion of dCas9 to the KRAB or SID4X repressors in eukaryotic systems can increase targeted repression (Jiang et al. 2013; Konermann et al. 2013; Perez-Pinera et al. 2013; Qi et al. 2013). Similar to transcriptional repression, dCas9 can also be fused to a transcriptional activator, such as VP64 in eukaryotic systems or the omega subunit of RNA polymerase in prokaryotic systems (Cheng et al. 2013; Jiang et al. 2013; Mali et al. 2013; Perez-Pinera et al. 2013; Qi et al. 2013). When guided to a promoter, these dCas9-activator fusions can efficiently recruit RNA polymerase and activate transcription of genes of interest.

The programmable DNA binding activity of dCas9 has been exploited even further. For instance, a fluorescently tagged dCas9 can be guided to specific genetic loci in live cells, allowing the spatiotemporal dynamics of specific sequences within the chromatin to be observed (Chen et al. 2013). Additionally, dCas9 has also been used to purify specific DNA sequences from live cells, in an enhanced form of chromatin immunoprecipitation (enCHiP) (Fujita and Fujii 2015). Cas9-mediated enCHiP has allowed the identification of previously unknown proteins that associate with specific DNA sequences in mammalian chromosomes (Fujita and Fujii 2013, 2014). Furthermore, fusion of dCas9 to the human acetyltransferase p300 allows the site-specific acetylation of histone H3 on lysine 27 (Hilton et al. 2015). This facilitates the activation of genes at enhancer sites distal to the targeted gene and also allows heritable epigenetic changes to be passed into a population (Hilton et al. 2015). Future Cas9 technologies may use other effector proteins to drive sequence-specific epigenetic modifications, such as DNA and/or histone methylation.

One of the most powerful attributes of the Cas9 system is the ability to be multiplexed to distinct targets within the same cell (Cheng et al. 2013; Cong et al. 2013; Ryan and Cate 2014); see Chapter 3 Introduction: Characterization of Cas9–Guide RNA Orthologs (Braff et al. 2016). In fact, the simultaneous utilization of Cas9 orthologs from distinct species has allowed the generation of mutations, as well as transcriptional activation and repression to occur within the same cell (Esvelt et al. 2013). Such methods lay the foundation for the engineering of incredibly detailed genetic circuits or to intricately probe genetic networks. In theory, the multiplexing capacity of Cas9 could be used to generate double- and triple-mutant libraries, facilitating the study of redundant systems and more easily exploring complex genetic circuits.

Despite the unprecedented utility and efficiency of the Cas9-dependent tools that have been created, one nontrivial challenge facing these technologies is off-target effects. Outside of a seed sequence located up to 12 bases proximal to the PAM, Cas9 can tolerate a range of mismatches, allowing it to bind and cleave sequences that are not the exact target (Jinek et al. 2012; Cradick et al. 2013; Pattanayak et al. 2013; Lin et al. 2014b). To prevent nontarget interactions, a number of databases have been developed (such as E-Crisp, Off-Spotter, and CRISPRdirect) that allow researchers to design gRNAs with

 Cite this introduction as *Cold Spring Harb Protoc*; doi:10.1101/pdb.top088849

optimized targeting and few to no off-target possibilities (Heigwer et al. 2014; Naito et al. 2015; Pliatsika and Rigoutsos 2015). However, such optimized Cas9 targeting can still be somewhat imperfect.

One method to drastically reduce off-target effects involves guiding Cas9 nickases to offset sites on the opposite strands, flanking the target, and creating a pair of ssDNA nicks (Mali et al. 2013; Ran et al. 2013). In conjunction with a donor construct containing homology with the sequences adjacent to those that have been cleaved, this method allows very high specificity of gene replacement at the site flanked by the offset nicks. Off-target effects are significantly limited, as the likelihood of nicked pairs at sites other than the desired sequence is extremely low (Mali et al. 2013; Ran et al. 2013). Furthermore, ssDNA nicks are easily repaired by the cell with almost undetectable levels of mutation. Thus, even if a single Cas9 nickase cleaves an off target site, the likelihood of a detrimental effect is limited. Cas9 has also been recently engineered to contain a photocaged lysine, rendering the protein catalytically inactive until stimulated with UV light, allowing it to become active and capable of cleaving DNA targets (Hemphill et al. 2015). Although still in infancy, such approaches will allow a fine-tuning of the regulation of Cas9 catalytic activity. These methods to overcome the potential off-target and other undesired effects of Cas9 will greatly increase the utility and acceptance of this technology, not only in a research setting, but also in therapeutic and clinical applications.

Cas9 technologies hold promise for use in mediating gene therapy, although numerous significant hurdles and questions remain. Although delivery (described in Chapter 5 Protocol 1: Adeno-Associated Virus–Mediated Delivery of CRISPR–Cas Systems for Genome Engineering in Mammalian Cells [Gaj and Schaffer 2016]) is a major roadblock, supplying Cas9, specific gRNAs, and repair constructs may allow the treatment of defined genetic disorders, by introducing or removing genetic information. Although large in size, Cas9 may be packaged into adeno- and lentiviral vectors (Shalem et al. 2014; Ran et al. 2015), but recent studies have also showed that Cas9 in complex with gRNAs can enter cells directly using lipid-based transfection techniques, fusion to cell-penetrating peptides, and nanoparticle delivery (Ramakrishna et al. 2014; McNeer et al. 2015; Zuris et al. 2015). Furthermore, the study of various Cas9 variants from different species may reveal a minimally sized Cas9 enzyme that retains programmable DNA binding and cleavage function (Jinek et al. 2014; Ran et al. 2015). An additional layer of security in delivery has also been successfully used whereby Cas9 is controlled by cell-specific promoters, allowing its activity to be limited to very specific cell types, such as neurons (Swiech et al. 2015). Further approaches using optogenetics have allowed the regulation of dCas9-mediated gene activation only in response to light stimulation (Konermann et al. 2013; Hemphill et al. 2015; Nihongaki et al. 2015).

The pathway toward translational uses of Cas9-directed repair has been exemplified recently in a number of systems. For instance, a common mutation in the CFTR locus that contributes to cystic fibrosis was repaired by Cas9 in primary human intestinal cells (Schwank et al. 2013). Similarly, in human induced pluripotent stem cells (iPSCs) derived from a myeloproliferative neoplasm, Cas9 was used to repair the oncogenic mutation (Smith et al. 2015), and mutations in the *crygc* gene that is responsible for cataracts were repaired in mouse zygotes and spermatogonial stem cells (Ren et al. 2013; Wu et al. 2015). Additionally, HIV proviruses have been removed from infected cells using Cas9-directed cleavage, and hepatitis B and hepatitis C viruses have been targeted, perhaps providing a framework for future antiviral therapeutics (Hu et al. 2014; Lin et al. 2014a; Kennedy et al. 2015; Liao et al. 2015; Price et al. 2015). Such repair has not been limited to tissue culture studies ex vivo. In mice (Ren et al. 2013), the *Fah* mutation, which induces tyrosinemia, and recently a *cftr* mutation in a mouse model of cystic fibrosis were both successfully corrected through Cas9-mediated repair (Yin et al. 2014; McNeer et al. 2015). Although proofs of concept, these groundbreaking studies highlight the therapeutic potential of emerging Cas9 technologies in treating genetic disorders.

FUTURE DIRECTIONS

From their first identification as unique genetic elements to the elucidation of their function as a prokaryotic adaptive immune system, CRISPR–Cas systems have been one of the most exciting fields in biology. Being able to exploit these systems for biotechnological purposes serves to emphasize the power that the study of seemingly "basic" biological mechanisms can have on extremely far reaching biotechnological and clinical applications. Already, Cas9-mediated engineering has been used throughout multiple fields and is rapidly changing the face of eukaryotic genetics.

Continued study of natural CRISPR–Cas systems, both in their canonical function as restriction systems against nucleic acids and in their alternative roles in bacterial physiology, will provide further insights into how these systems can be exploited for bioengineering applications. As more Cas9 orthologs are analyzed, these variants will allow researchers to further understand the structural and sequence requirements that determine PAM specificity, crRNA sequence requirements, and DNA binding stringency, allowing Cas9 proteins to be engineered for increased specificity and efficacy. Likewise, there remain large, unanswered questions in the field of CRISPR–Cas biology that will certainly lead to the development of even more tools for molecular biology. Already, other Cas proteins have been predicted to have diverse and conserved functions. For example, Cas1 and Cas2 have been proposed to act as a toxin–antitoxin system, becoming autotoxic in the presence of bacteriophage infections that are not successfully controlled by the canonical nucleic acid targeting activity of CRISPR–Cas systems, perhaps by cleaving endogenous mRNA (Makarova et al. 2012). This second line of defense would prevent bacteriophages from replicating and subsequently infecting other cells but, if true, could also form the platform for a Cas2-based RNA interference technology. At the same time, continued study of how Cas1 and Cas2 act to integrate new sequences into the bacterial chromosome may further allow the generation of new technologies that are more efficient at mediating site-directed DNA integration. Thus, as we learn more about the functions of diverse Cas proteins, we will greatly expand our ability to develop novel molecular tools for interrogation of pressing biological mysteries.

The power of proteins that can be programmed to recognize specific sequences of DNA is enormous. Given the ease and accessibility of the Cas9 system, incredible progress has been made in developing this system for a plethora of purposes that have already left their mark on numerous disciplines from molecular biology to translational medicine. Cas9 has shaped and will continue to shape modern biology now and for the foreseeable future. The technological possibilities of Cas9 are seemingly endless and limited only by our creativity and imagination.

ACKNOWLEDGMENTS

Because of space constraints and the rapidly expanding field, we have undoubtedly omitted notable work in the field; we sincerely apologize to those authors whose research we did not cite. This work was supported by National Institutes of Health grant R01-AI11070 and a Burroughs Wellcome Fund Investigator in the Pathogenesis of Infectious Disease award to D.S.W.

REFERENCES

Anders C, Niewoehner O, Duerst A, Jinek M. 2014. Structural basis of PAM-dependent target DNA recognition by the Cas9 endonuclease. *Nature* **513:** 569–573.

Barrangou R. 2014. Cas9 targeting and the CRISPR revolution. *Science* **344:** 707–708.

Barrangou R. 2015. The roles of CRISPR–Cas systems in adaptive immunity and beyond. *Curr Opin Immunol* **32:** 36–41.

Barrangou R, Marraffini LA. 2014. CRISPR-Cas systems: Prokaryotes upgrade to adaptive immunity. *Mol Cell* **54:** 234–244.

Barrangou R, Fremaux C, Deveau H, Richards M, Boyaval P, Moineau S, Romero DA, Horvath P. 2007. CRISPR–provides acquired resistance against viruses in prokaryotes. *Science* **315:** 1709–1712.

Belhaj K, Chaparro-Garcia A, Kamoun S, Nekrasov V. 2013. Plant genome editing made easy: Targeted mutagenesis in model and crop plants using the CRISPR/Cas system. *Plant Methods* **9:** 39.

Bell CC, Magor GW, Gillinder KR, Perkins AC. 2014. A high-throughput screening strategy for detecting CRISPR–Cas9 induced mutations using next-generation sequencing. *BMC Genomics* **15:** 1002.

Bikard D, Marraffini LA. 2013. Control of gene expression by CRISPR–Cas systems. *F1000Prime Rep* **5:** 47.

Bikard D, Hatoum-Aslan A, Mucida D, Marraffini LA. 2012. CRISPR interference can prevent natural transformation and virulence acquisition during in vivo bacterial infection. *Cell Host Microbe* **12:** 177–186.

Bolotin A, Quinquis B, Sorokin A, Ehrlich SD. 2005. Clustered regularly interspaced short palindrome repeats (CRISPRs) have spacers of extrachromosomal origin. *Microbiology* **151:** 2551–2561.

Braff JL, Yaung SJ, Esvelt KM, Church GM. 2016. Characterization of Cas9–guide RNA orthologs. *Cold Spring Harb Protoc* doi: 10.1101/pdb.top086793.

 Cite this introduction as *Cold Spring Harb Protoc*; doi:10.1101/pdb.top088849

Briner AE, Barrangou R. 2016. Guide RNAs: A glimpse at the sequences that drive CRISPR–Cas systems. *Cold Spring Harb Protoc* doi: 10.1101/pdb.top090902.

Brouns SJ, Jore MM, Lundgren M, Westra ER, Slijkhuis RJ, Snijders AP, Dickman MJ, Makarova KS, Koonin EV, van der Oost J. 2008. Small CRISPR RNAs guide antiviral defense in prokaryotes. *Science* **321:** 960–964.

Cady KC, O'Toole GA. 2011. Non-identity-mediated CRISPR-bacteriophage interaction mediated via the Csy and Cas3 proteins. *J Bacteriol* **193:** 3433–3445.

Carte J, Wang R, Li H, Terns RM, Terns MP. 2008. Cas6 is an endoribonuclease that generates guide RNAs for invader defense in prokaryotes. *Genes Dev* **22:** 3489–3496.

Carte J, Pfister NT, Compton MM, Terns RM, Terns MP. 2010. Binding and cleavage of CRISPR RNA by Cas6. *RNA* **16:** 2181–2188.

Chen B, Gilbert LA, Cimini BA, Schnitzbauer J, Zhang W, Li GW, Park J, Blackburn EH, Weissman JS, Qi LS, et al. 2013. Dynamic imaging of genomic loci in living human cells by an optimized CRISPR/Cas system. *Cell* **155:** 1479–1491.

Cheng AW, Wang H, Yang H, Shi L, Katz Y, Theunissen TW, Rangarajan S, Shivalila CS, Dadon DB, Jaenisch R. 2013. Multiplexed activation of endogenous genes by CRISPR-on, an RNA-guided transcriptional activator system. *Cell Res* **23:** 1163–1171.

Cho SW, Kim S, Kim JM, Kim JS. 2013. Targeted genome engineering in human cells with the Cas9 RNA-guided endonuclease. *Nat Biotechnol* **31:** 230–232.

Chu VT, Weber T, Wefers B, Wurst W, Sander S, Rajewsky K, Kuhn R. 2015. Increasing the efficiency of homology-directed repair for CRISPR–Cas9-induced precise gene editing in mammalian cells. *Nat Biotechnol* **33:** 543–548.

Chylinski K, Le Rhun A, Charpentier E. 2013. The tracrRNA and Cas9 families of type II CRISPR–Cas immunity systems. *RNA Biol* **10:** 726–737.

Chylinski K, Makarova KS, Charpentier E, Koonin EV. 2014. Classification and evolution of type II CRISPR–Cas systems. *Nucleic Acids Res* **42:** 6091–6105.

Cong L, Ran FA, Cox D, Lin S, Barretto R, Habib N, Hsu PD, Wu X, Jiang W, Marraffini LA, et al. 2013. Multiplex genome engineering using CRISPR/Cas systems. *Science* **339:** 819–823.

Cradick TJ, Fine EJ, Antico CJ, Bao G. 2013. CRISPR/Cas9 systems targeting β-globin and *CCR5* genes have substantial off-target activity. *Nucleic Acids Res* **41:** 9584–9592.

Datsenko KA, Pougach K, Tikhonov A, Wanner BL, Severinov K, Semenova E. 2012. Molecular memory of prior infections activates the CRISPR/Cas adaptive bacterial immunity system. *Nat Commun* **3:** 945.

Deltcheva E, Chylinski K, Sharma CM, Gonzales K, Chao Y, Pirzada ZA, Eckert MR, Vogel J, Charpentier E. 2011. CRISPR RNA maturation by trans-encoded small RNA and host factor RNase III. *Nature* **471:** 602–607.

Deveau H, Barrangou R, Garneau JE, Labonté J, Fremaux C, Boyaval P, Romero DA, Horvath P, Moineau S. 2008. Phage response to CRISPR-encoded resistance in *Streptococcus thermophilus. J Bacteriol* **190:** 1390–1400.

DiCarlo JE, Norville JE, Mali P, Rios X, Aach J, Church GM. 2013. Genome engineering in *Saccharomyces cerevisiae* using CRISPR–Cas systems. *Nucleic Acids Res* **41:** 4336–4343.

Díez-Villaseñor C, Guzmán NM, Almendros C, García-Martínez J, Mojica FJM. 2013. CRISPR-spacer integration reporter plasmids reveal distinct genuine acquisition specificities among CRISPR-Cas I-E variants of *Escherichia coli. RNA Biol* **10:** 792–802.

Doudna JA, Charpentier E. 2014. Genome editing. The new frontier of genome engineering with CRISPR-Cas9. *Science* **346:** 1258096.

Du D, Qi LS. 2016. CRISPR technology for genome activation and repression in mammalian cells. *Cold Spring Harb Protoc* doi: 10.1101/pdb.prot090175.

Esvelt KM, Mali P, Braff JL, Moosburner M, Yaung SJ, Church GM. 2013. Orthogonal Cas9 proteins for RNA-guided gene regulation and editing. *Nat Methods* **10:** 1116–1121.

Flowers GP, Timberlake AT, McLean KC, Monaghan JR, Crews CM. 2014. Highly efficient targeted mutagenesis in axolotl using Cas9 RNA-guided nuclease. *Development* **141:** 2165–2171.

Fonfara I, Le Rhun A, Chylinski K, Makarova KS, Lecrivain AL, Bzdrenga J, Koonin EV, Charpentier E. 2014. Phylogeny of Cas9 determines functional exchangeability of dual-RNA and Cas9 among orthologous type II CRISPR–Cas systems. *Nucleic Acids Res* **42:** 2577–2590.

Fujita T, Fujii H. 2013. Efficient isolation of specific genomic regions and identification of associated proteins by engineered DNA-binding molecule-mediated chromatin immunoprecipitation (enChIP) using CRISPR. *Biochem Biophys Res Commun* **439:** 132–136.

Fujita T, Fujii H. 2014. Identification of proteins associated with an IFNγ-responsive promoter by a retroviral expression system for enChIP using CRISPR. *PLoS One* **9:** e103084.

Fujita T, Fujii H. 2015. Isolation of specific genomic regions and identification of associated molecules by engineered DNA-binding molecule-mediated chromatin immunoprecipitation (enChIP) using CRISPR. *Methods Mol Biol* **1288:** 43–52.

Gaj T, Schaffer DV. 2016. Adeno-associated virus–mediated delivery of CRISPR–Cas systems for genome engineering in mammalian cells. *Cold Spring Harb Protoc* doi: 10.1101/pdb.prot086868.

Garneau JE, Dupuis ME, Villion M, Romero DA, Barrangou R, Boyaval P, Fremaux C, Horvath P, Magadan AH, Moineau S. 2010. The CRISPR/Cas bacterial immune system cleaves bacteriophage and plasmid DNA. *Nature* **468:** 67–71.

Garside EL, Schellenberg MJ, Gesner EM, Bonanno JB, Sauder JM, Burley SK, Almo SC, Mehta G, MacMillan AM. 2012. Cas5d processes pre-crRNA and is a member of a larger family of CRISPR RNA endonucleases. *RNA* **18:** 2020–2028.

Gasiunas G, Barrangou R, Horvath P, Siksnys V. 2012. Cas9-crRNA ribonucleoprotein complex mediates specific DNA cleavage for adaptive immunity in bacteria. *Proc Natl Acad Sci* **109:** E2579–E2586.

Gesner EM, Schellenberg MJ, Garside EL, George MM, MacMillan AM. 2011. Recognition and maturation of effector RNAs in a CRISPR interference pathway. *Nat Struct Mol Biol* **18:** 688–692.

Goldberg GW, Jiang W, Bikard D, Marraffini LA. 2014. Conditional tolerance of temperate phages via transcription-dependent CRISPR–Cas targeting. *Nature* **514:** 633–637.

Gratz SJ, Cummings AM, Nguyen JN, Hamm DC, Donohue LK, Harrison MM, Wildonger J, O'Connor-Giles KM. 2013. Genome engineering of *Drosophila* with the CRISPR RNA-guided Cas9 nuclease. *Genetics* **194:** 1029–1035.

Gunderson FF, Cianciotto NP. 2013. The CRISPR-associated gene *cas2* of *Legionella pneumophila* is required for intracellular infection of amoebae. *mBio* **4:** e00074–e00013.

Gunderson FF, Mallama CA, Fairbairn SG, Cianciotto NP. 2015. Nuclease activity of *Legionella pneumophila* Cas2 promotes intracellular infection of amoebal host cells. *Infect Immun* **83:** 1008–1018.

Haft DH, Selengut J, Mongodin EF, Nelson KE. 2005. A guild of 45 CRISPR-associated (Cas) protein families and multiple CRISPR/Cas subtypes exist in prokaryotic genomes. *PLoS Comput Biol* **1:** e60.

Hale CR, Zhao P, Olson S, Duff MO, Graveley BR, Wells L, Terns RM, Terns MP. 2009. RNA-guided RNA cleavage by a CRISPR RNA–Cas protein complex. *Cell* **139:** 945–956.

Hatoum-Aslan A, Maniv I, Marraffini LA. 2011. Mature clustered, regularly interspaced, short palindromic repeats RNA (crRNA) length is measured by a ruler mechanism anchored at the precursor processing site. *Proc Natl Acad Sci* **108:** 21218–21222.

Haurwitz RE, Jinek M, Wiedenheft B, Zhou K, Doudna JA. 2010. Sequence- and structure-specific RNA processing by a CRISPR endonuclease. *Science* **329:** 1355–1358.

Haurwitz RE, Sternberg SH, Doudna JA. 2012. Csy4 relies on an unusual catalytic dyad to position and cleave CRISPR RNA. *EMBO J* **31:** 2824–2832.

Heidrich N, Vogel J. 2013. CRISPRs extending their reach: Prokaryotic RNAi protein Cas9 recruited for gene regulation. *EMBO J* **32:** 1802–1804.

Heigwer F, Kerr G, Boutros M. 2014. E-CRISP: Fast CRISPR target site identification. *Nat Methods* **11:** 122–123.

Heler R, Marraffini LA, Bikard D. 2014. Adapting to new threats: The generation of memory by CRISPR–Cas immune systems. *Mol Microbiol* **93:** 1–9.

Heler R, Samai P, Modell JW, Weiner C, Goldberg GW, Bikard D, Marraffini LA. 2015. Cas9 specifies functional viral targets during CRISPR–Cas adaptation. *Nature* **519:** 199–202.

Hemphill J, Borchardt EK, Brown K, Asokan A, Deiters A. 2015. Optical control of CRISPR/Cas9 gene editing. *J Am Chem Soc* **137:** 5642–5645.

Henao-Mejia J, Williams A, Rongvaux A, Stein J, Hughes C, Flavell RA. 2016. Generation of genetically modified mice using the CRISPR-

Cas9 genome-editing system. *Cold Spring Harb Protoc* doi: 10.1101/pdb.prot090704.

Hilton IB, D'Ippolito AM, Vockley CM, Thakore PI, Crawford GE, Reddy TE, Gersbach CA. 2015. Epigenome editing by a CRISPR–Cas9-based acetyltransferase activates genes from promoters and enhancers. *Nat Biotechnol* **33:** 510–517.

Hou Z, Zhang Y, Propson NE, Howden SE, Chu LF, Sontheimer EJ, Thomson JA. 2013. Efficient genome engineering in human pluripotent stem cells using Cas9 from *Neisseria meningitidis*. *Proc Natl Acad Sci* **110:** 15644–15649.

Housden BE, Perrimon N. 2016. Cas9-mediated genome engineering in *Drosophila melanogaster*. *Cold Spring Harb Protoc* doi: 10.1101/pdb.top086843.

Hrle A, Su AAH, Ebert J, Benda C, Randau L, Conti E. 2013. Structure and RNA-binding properties of the Type III-A CRISPR-associated protein Csm3. *RNA Biol* **10:** 1670–1678.

Hu W, Kaminski R, Yang F, Zhang Y, Cosentino L, Li F, Luo B, Alvarez-Carbonell D, Garcia-Mesa Y, Karn J, et al. 2014. RNA-directed gene editing specifically eradicates latent and prevents new HIV-1 infection. *Proc Natl Acad Sci* **111:** 11461–11466.

Hwang WY, Fu Y, Reyon D, Maeder ML, Tsai SQ, Sander JD, Peterson RT, Yeh JR, Joung JK. 2013. Efficient genome editing in zebrafish using a CRISPR–Cas system. *Nat Biotechnol* **31:** 227–229.

Ishino Y, Shinagawa H, Makino K, Amemura M, Nakata A. 1987. Nucleotide sequence of the iap gene, responsible for alkaline phosphatase isozyme conversion in *Escherichia coli*, and identification of the gene product. *J Bacteriol* **169:** 5429–5433.

Jansen R, Embden JD, Gaastra W, Schouls LM. 2002. Identification of genes that are associated with DNA repeats in prokaryotes. *Mol Microbiol* **43:** 1565–1575.

Jiang W, Bikard D, Cox D, Zhang F, Marraffini LA. 2013. RNA-guided editing of bacterial genomes using CRISPR–Cas systems. *Nat Biotechnol* **31:** 233–239.

Jinek M, Chylinski K, Fonfara I, Hauer M, Doudna JA, Charpentier E. 2012. A programmable dual-RNA-guided DNA endonuclease in adaptive bacterial immunity. *Science* **337:** 816–821.

Jinek M, East A, Cheng A, Lin S, Ma E, Doudna J. 2013. RNA-programmed genome editing in human cells. *eLife* **2:** e00471.

Jinek M, Jiang F, Taylor DW, Sternberg SH, Kaya E, Ma E, Anders C, Hauer M, Zhou K, Lin S, et al. 2014. Structures of Cas9 endonucleases reveal RNA-mediated conformational activation. *Science* **343:** 1247997.

Jones CL, Sampson TR, Nakaya HI, Pulendran B, Weiss DS. 2012. Repression of bacterial lipoprotein production by *Francisella novicida* facilitates evasion of innate immune recognition. *Cell Microbiol* **14:** 1531–1543.

Jore MM, Lundgren M, van Duijn E, Bultema JB, Westra ER, Waghmare SP, Wiedenheft B, Pul Ü, Wurm R, Wagner R, et al. 2011. Structural basis for CRISPR RNA-guided DNA recognition by Cascade. *Nat Struct Mol Biol* **18:** 529–536.

Kennedy EM, Bassit LC, Mueller H, Kornepati AV, Bogerd HP, Nie T, Chatterjee P, Javanbakht H, Schinazi RF, Cullen BR. 2015. Suppression of hepatitis B virus DNA accumulation in chronically infected cells using a bacterial CRISPR/Cas RNA-guided DNA endonuclease. *Virology* **476:** 196–205.

Konermann S, Brigham MD, Trevino AE, Hsu PD, Heidenreich M, Cong L, Platt RJ, Scott DA, Church GM, Zhang F. 2013. Optical control of mammalian endogenous transcription and epigenetic states. *Nature* **500:** 472–476.

Koo Y, Ka D, Kim E-J, Suh N, Bae E. 2013. Conservation and variability in the structure and function of the Cas5d endoribonuclease in the CRISPR-mediated microbial immune system. *J Mol Biol* **425:** 3799–3810.

Koonin EV, Krupovic M. 2015. Evolution of adaptive immunity from transposable elements combined with innate immune systems. *Nat Rev Genet* **16:** 184–192.

Krupovic M, Makarova KS, Forterre P, Prangishvili D, Koonin EV. 2014. Casposons: A new superfamily of self-synthesizing DNA transposons at the origin of prokaryotic CRISPR–Cas immunity. *BMC Biol* **12:** 36.

Liao HK, Gu Y, Diaz A, Marlett J, Takahashi Y, Li M, Suzuki K, Xu R, Hishida T, Chang CJ, et al. 2015. Use of the CRISPR/Cas9 system as an intracellular defense against HIV-1 infection in human cells. *Nat Commun* **6:** 6413.

Lin SR, Yang HC, Kuo YT, Liu CJ, Yang TY, Sung KC, Lin YY, Wang HY, Wang CC, Shen YC, et al. 2014a. The CRISPR/Cas9 system facilitates clearance of the intrahepatic HBV templates in vivo. *Mol Ther Nucleic Acids* **3:** e186.

Lin Y, Cradick TJ, Brown MT, Deshmukh H, Ranjan P, Sarode N, Wile BM, Vertino PM, Stewart FJ, Bao G. 2014b. CRISPR/Cas9 systems have off-target activity with insertions or deletions between target DNA and guide RNA sequences. *Nucleic Acids Res* **42:** 7473–7485.

Lintner NG, Kerou M, Brumfield SK, Graham S, Liu H, Naismith JH, Sdano M, Peng N, She Q, Copié V, et al. 2011. Structural and functional characterization of an archaeal clustered regularly interspaced short palindromic repeat (CRISPR)-associated complex for antiviral defense (CASCADE). *J Biol Chem* **286:** 21643–21656.

Lo TW, Pickle CS, Lin S, Ralston EJ, Gurling M, Schartner CM, Bian Q, Doudna JA, Meyer BJ. 2013. Precise and heritable genome editing in evolutionarily diverse nematodes using TALENs and CRISPR/Cas9 to engineer insertions and deletions. *Genetics* **195:** 331–348.

Louwen R, Horst-Kreft D, de Boer AG, van der Graaf L, de Knegt G, Hamersma M, Heikema AP, Timms AR, Jacobs BC, Wagenaar JA, et al. 2013. A novel link between *Campylobacter jejuni* bacteriophage defence, virulence and Guillain–Barre syndrome. *Eur J Clin Microbiol Infect Dis* **32:** 207–226.

Makarova KS, Grishin NV, Shabalina SA, Wolf YI, Koonin EV. 2006. A putative RNA-interference-based immune system in prokaryotes: Computational analysis of the predicted enzymatic machinery, functional analogies with eukaryotic RNAi, and hypothetical mechanisms of action. *Biol Direct* **1:** 7.

Makarova KS, Haft DH, Barrangou R, Brouns SJ, Charpentier E, Horvath P, Moineau S, Mojica FJ, Wolf YI, Yakunin AF, et al. 2011. Evolution and classification of the CRISPR-Cas systems. *Nat Rev Microbiol* **9:** 467–477.

Makarova KS, Anantharaman V, Aravind L, Koonin EV. 2012. Live virus-free or die: Coupling of antivirus immunity and programmed suicide or dormancy in prokaryotes. *Biol Direct* **7:** 40.

Makarova KS, Wolf YI, Koonin EV. 2013. Comparative genomics of defense systems in archaea and bacteria. *Nucleic Acids Res* **41:** 4360–4377.

Mali P, Aach J, Stranges PB, Esvelt KM, Moosburner M, Kosuri S, Yang L, Church GM. 2013. CAS9 transcriptional activators for target specificity screening and paired nickases for cooperative genome engineering. *Nat Biotechnol* **31:** 833–838.

Marraffini LA, Sontheimer EJ. 2008. CRISPR interference limits horizontal gene transfer in staphylococci by targeting DNA. *Science* **322:** 1843–1845.

Marraffini LA, Sontheimer EJ. 2010. Self vs. non-self discrimination during CRISPR RNA-directed immunity. *Nature* **463:** 568–571.

Maruyama T, Dougan SK, Truttmann MC, Bilate AM, Ingram JR, Ploegh HL. 2015. Increasing the efficiency of precise genome editing with CRISPR–Cas9 by inhibition of nonhomologous end joining. *Nat Biotechnol* **33:** 538–542.

McNeer NA, Anandalingam K, Fields RJ, Caputo C, Kopic S, Gupta A, Quijano E, Polikoff L, Kong Y, Bahal R, et al. 2015. Nanoparticles that deliver triplex-forming peptide nucleic acid molecules correct F508del CFTR in airway epithelium. *Nat Commun* **6:** 6952.

Miyaoka Y, Chan AH, Conklin BR. 2016. Detecting single-nucleotide substitutions induced by genome editing. *Cold Spring Harb Protoc* doi: 10.1101/pdb.top090845.

Mojica FJ, Diez-Villasenor C, Garcia-Martinez J, Soria E. 2005. Intervening sequences of regularly spaced prokaryotic repeats derive from foreign genetic elements. *J Mol Evol* **60:** 174–182.

Mojica FJM, Díez-Villaseñor C, García-Martínez J, Almendros C. 2009. Short motif sequences determine the targets of the prokaryotic CRISPR defence system. *Microbiology* **155:** 733–740.

Naito Y, Hino K, Bono H, Ui-Tei K. 2015. CRISPRdirect: Software for designing CRISPR/Cas guide RNA with reduced off-target sites. *Bioinformatics* **31:** 1120–1123.

Nakayama T, Fish MB, Fisher M, Oomen-Hajagos J, Thomsen GH, Grainger RM. 2013. Simple and efficient CRISPR/Cas9-mediated targeted mutagenesis in *Xenopus tropicalis*. *Genesis* **51:** 835–843.

Nam KH, Haitjema C, Liu X, Ding F, Wang H, DeLisa MP, Ke A. 2012. Cas5d protein processes pre-crRNA and assembles into a Cascade-like interference complex in Subtype I-C/Dvulg CRISPR-Cas system. *Structure* **20:** 1574–1584.

Nekrasov V, Staskawicz B, Weigel D, Jones JD, Kamoun S. 2013. Targeted mutagenesis in the model plant *Nicotiana benthamiana* using Cas9 RNA-guided endonuclease. *Nat Biotechnol* **31:** 691–693.

Niewoehner O, Jinek M, Doudna JA. 2014. Evolution of CRISPR RNA recognition and processing by Cas6 endonucleases. *Nucleic Acids Res* **42:** 1341–1353.

Nihongaki Y, Yamamoto S, Kawano F, Suzuki H, Sato M. 2015. CRISPR–Cas9-based photoactivatable transcription system. *Chem Biol* **22:** 169–174.

Nishimasu H, Ran FA, Hsu PD, Konermann S, Shehata SI, Dohmae N, Ishitani R, Zhang F, Nureki O. 2014. Crystal structure of Cas9 in complex with guide RNA and target DNA. *Cell* **156:** 935–949.

Nunez JK, Kranzusch PJ, Noeske J, Wright AV, Davies CW, Doudna JA. 2014. Cas1–Cas2 complex formation mediates spacer acquisition during CRISPR–Cas adaptive immunity. *Nat Struct Mol Biol* **21:** 528–534.

Nunez JK, Lee AS, Engelman A, Doudna JA. 2015. Integrase-mediated spacer acquisition during CRISPR–Cas adaptive immunity. *Nature* **519:** 193–198.

Pattanayak V, Lin S, Guilinger JP, Ma E, Doudna JA, Liu DR. 2013. High-throughput profiling of off-target DNA cleavage reveals RNA-programmed Cas9 nuclease specificity. *Nat Biotechnol* **31:** 839–843.

Peng W, Feng M, Feng X, Liang YX, She Q. 2015. An archaeal CRISPR type III-B system exhibiting distinctive RNA targeting features and mediating dual RNA and DNA interference. *Nucleic Acids Res* **43:** 406–417.

Pennisi E. 2013. The CRISPR craze. *Science* **341:** 833–836.

Perez-Pinera P, Kocak DD, Vockley CM, Adler AF, Kabadi AM, Polstein LR, Thakore PI, Glass KA, Ousterout DG, Leong KW, et al. 2013. RNA-guided gene activation by CRISPR–Cas9-based transcription factors. *Nat Methods* **10:** 973–976.

Plagens A, Tjaden B, Hagemann A, Randau L, Hensel R. 2012. Characterization of the CRISPR/Cas subtype I-A system of the hyperthermophilic crenarchaeon *Thermoproteus tenax. J Bacteriol* **194:** 2491–2500.

Plagens A, Richter H, Charpentier E, Randau L. 2015. DNA and RNA interference mechanisms by CRISPR–Cas surveillance complexes. *FEMS Microbiol Rev* **39:** 442–463.

Pliatsika V, Rigoutsos I. 2015. "Off-Spotter": Very fast and exhaustive enumeration of genomic lookalikes for designing CRISPR/Cas guide RNAs. *Biol Direct* **10:** 4.

Price AA, Sampson TR, Ratner HK, Grakoui A, Weiss DS. 2015. Cas9-mediated targeting of viral RNA in eukaryotic cells. *Proc Natl Acad Sci* **112:** 6164–6169.

Qi LS, Larson MH, Gilbert LA, Doudna JA, Weissman JS, Arkin AP, Lim WA. 2013. Repurposing CRISPR as an RNA-guided platform for sequence-specific control of gene expression. *Cell* **152:** 1173–1183.

Ramakrishna S, Kwaku Dad AB, Beloor J, Gopalappa R, Lee SK, Kim H. 2014. Gene disruption by cell-penetrating peptide-mediated delivery of Cas9 protein and guide RNA. *Genome Res* **24:** 1020–1027.

Ran FA, Hsu PD, Lin CY, Gootenberg JS, Konermann S, Trevino AE, Scott DA, Inoue A, Matoba S, Zhang Y, et al. 2013. Double nicking by RNA-guided CRISPR Cas9 for enhanced genome editing specificity. *Cell* **154:** 1380–1389.

Ran FA, Cong L, Yan WX, Scott DA, Gootenberg JS, Kriz AJ, Zetsche B, Shalem O, Wu X, Makarova KS, et al. 2015. In vivo genome editing using *Staphylococcus aureus* Cas9. *Nature* **520:** 186–191.

Rath D, Amlinger L, Rath A, Lundgren M. 2015. The CRISPR-Cas immune system: Biology, mechanisms and applications. *Biochimie* **117:** 119–128.

Ratner HK, Sampson TR, Weiss DS. 2015. I can see CRISPR now, even when phage are gone: A view on alternative CRISPR–Cas functions from the prokaryotic envelope. *Curr Opin Infect Dis* **28:** 267–274.

Reeks J, Naismith James H, White Malcolm F. 2013. CRISPR interference: A structural perspective. *Biochem J* **453:** 155–166.

Ren X, Sun J, Housden BE, Hu Y, Roesel C, Lin S, Liu LP, Yang Z, Mao D, Sun L, et al. 2013. Optimized gene editing technology for *Drosophila melanogaster* using germ line-specific Cas9. *Proc Natl Acad Sci* **110:** 19012–19017.

Rouillon C, Zhou M, Zhang J, Politis A, Beilsten-Edmands V, Cannone G, Graham S, Robinson CV, Spagnolo L, White MF. 2013. Structure of the CRISPR interference complex CSM reveals key similarities with cascade. *Mol Cell* **52:** 124–134.

Ryan OW, Cate JH. 2014. Multiplex engineering of industrial yeast genomes using CRISPRm. *Methods Enzymol* **546:** 473–489.

Ryan OW, Poddar S, Cate JHD. 2016. CRISPR–Cas9 genome engineering in *Saccharomyces cerevisiae* cells. *Cold Spring Harb Protoc* doi: 10.1101/pdb.prot086827.

Samai P, Pyenson N, Jiang W, Goldberg GW, Hatoum-Aslan A, Marraffini LA. 2015. Co-transcriptional DNA and RNA cleavage during type III CRISPR-Cas immunity. *Cell* **161:** 1164–1174.

Sampson TR, Weiss DS. 2013. Cas9-dependent endogenous gene regulation is required for bacterial virulence. *Biochem Soc Trans* **41:** 1407–1411.

Sampson TR, Weiss DS. 2014. Exploiting CRISPR/Cas systems for biotechnology. *Bioessays* **36:** 34–38.

Sampson TR, Saroj SD, Llewellyn AC, Tzeng YL, Weiss DS. 2013. A CRISPR/Cas system mediates bacterial innate immune evasion and virulence. *Nature* **497:** 254–257.

Sampson TR, Napier BA, Schroeder MR, Louwen R, Zhao J, Chin CY, Ratner HK, Llewellyn AC, Jones CL, Laroui H, et al. 2014. A CRISPR–Cas system enhances envelope integrity mediating antibiotic resistance and inflammasome evasion. *Proc Natl Acad Sci* **111:** 11163–11168.

Sashital DG, Jinek M, Doudna JA. 2011. An RNA-induced conformational change required for CRISPR RNA cleavage by the endoribonuclease Cse3. *Nat Struct Mol Biol* **18:** 680–687.

Schwank G, Koo BK, Sasselli V, Dekkers JF, Heo I, Demircan T, Sasaki N, Boymans S, Cuppen E, van der Ent CK, et al. 2013. Functional repair of CFTR by CRISPR/Cas9 in intestinal stem cell organoids of cystic fibrosis patients. *Cell Stem Cell* **13:** 653–658.

Shalem O, Sanjana NE, Hartenian E, Shi X, Scott DA, Mikkelsen TS, Heckl D, Ebert BL, Root DE, Doench JG, et al. 2014. Genome-scale CRISPR–Cas9 knockout screening in human cells. *Science* **343:** 84–87.

Smith C, Abalde-Atristain L, He C, Brodsky BR, Braunstein EM, Chaudhari P, Jang YY, Cheng L, Ye Z. 2015. Efficient and allele-specific genome editing of disease loci in human iPSCs. *Mol Ther* **23:** 570–577.

Smith C, Ye Z, Cheng L. 2016. A method for genome editing in human pluripotent stem cells. *Cold Spring Harb Protoc* doi: 10.1101/pdb.prot090217.

Spilman M, Cocozaki A, Hale C, Shao Y, Ramia N, Terns R, Terns M, Li H, Stagg S. 2013. Structure of an RNA silencing complex of the CRISPR–Cas immune system. *Mol Cell* **52:** 146–152.

Staals RH, Agari Y, Maki-Yonekura S, Zhu Y, Taylor DW, van Duijn E, Barendregt A, Vlot M, Koehorst JJ, Sakamoto K, et al. 2013. Structure and activity of the RNA-targeting type III-B CRISPR–Cas complex of *Thermus thermophilus. Mol Cell* **52:** 135–145.

Staals RH, Zhu Y, Taylor DW, Kornfeld JE, Sharma K, Barendregt A, Koehorst JJ, Vlot M, Neupane N, Varossieau K, et al. 2014. RNA targeting by the type III-A CRISPR–Cas Csm complex of *Thermus thermophilus. Mol Cell* **56:** 518–530.

Stern A, Keren L, Wurtzel O, Amitai G, Sorek R. 2010. Self-targeting by CRISPR: Gene regulation or autoimmunity? *Trends Genet* **26:** 335–340.

Sternberg SH, Haurwitz RE, Doudna JA. 2012. Mechanism of substrate selection by a highly specific CRISPR endoribonuclease. *RNA* **18:** 661–672.

Sternberg SH, Redding S, Jinek M, Greene EC, Doudna JA. 2014. DNA interrogation by the CRISPR RNA-guided endonuclease Cas9. *Nature* **507:** 62–67.

Swarts DC, Mosterd C, van Passel MWJ, Brouns SJJ. 2012. CRISPR interference directs strand specific spacer acquisition. *PLoS One* **7:** e35888.

Swiech L, Heidenreich M, Banerjee A, Habib N, Li Y, Trombetta J, Sur M, Zhang F. 2015. In vivo interrogation of gene function in the mammalian brain using CRISPR–Cas9. *Nat Biotechnol* **33:** 102–106.

van der Oost J, Westra ER, Jackson RN, Wiedenheft B. 2014. Unravelling the structural and mechanistic basis of CRISPR–Cas systems. *Nat Rev Microbiol* **12:** 479–492.

van Duijn E, Barbu IM, Barendregt A, Jore MM, Wiedenheft B, Lundgren M, Westra ER, Brouns SJJ, Doudna JA, van der Oost J, et al. 2012. Native tandem and ion mobility mass spectrometry highlight structural and modular similarities in clustered-regularly-interspaced short-palindromic-repeats (CRISPR)-associated protein complexes from *Escherichia coli* and *Pseudomonas aeruginosa. Mol Cell Proteomics* **11:** 1430–1441.

Veesenmeyer JL, Andersen AW, Lu X, Hussa EA, Murfin KE, Chaston JM, Dillman AR, Wassarman KM, Sternberg PW, Goodrich-Blair H. 2014.

NilD CRISPR RNA contributes to *Xenorhabdus nematophila* colonization of symbiotic host nematodes. *Mol Microbiol* **93:** 1026–1042.

Vejnar CE, Moreno-Mateos MA, Cifuentes D, Bazzini AA, Giraldez AJ. 2016. Optimized CRISPR–Cas9 system for genome editing in zebrafish. *Cold Spring Harb Protoc* doi: 10.1101/pdb.prot086850.

Vestergaard G, Garrett RA, Shah SA. 2014. CRISPR adaptive immune systems of Archaea. *RNA Biol* **11:** 156–167.

Viswanathan P, Murphy K, Julien B, Garza AG, Kroos L. 2007. Regulation of dev, an operon that includes genes essential for *Myxococcus xanthus* development and CRISPR-associated genes and repeats. *J Bacteriol* **189:** 3738–3750.

Wallace RA, Black WP, Yang X, Yang Z. 2014. A CRISPR with roles in *Myxococcus xanthus* development and exopolysaccharide production. *J Bacteriol* **196:** 4036–4043.

Wang R, Preamplume G, Terns MP, Terns RM, Li H. 2011. Interaction of the Cas6 riboendonuclease with CRISPR RNAs: Recognition and cleavage. *Structure* **19:** 257–264.

Wang H, Yang H, Shivalila CS, Dawlaty MM, Cheng AW, Zhang F, Jaenisch R. 2013. One-step generation of mice carrying mutations in multiple genes by CRISPR/Cas-mediated genome engineering. *Cell* **153:** 910–918.

Wang T, Wei JJ, Sabatini DM, Lander ES. 2014. Genetic screens in human cells using the CRISPR–Cas9 system. *Science* **343:** 80–84.

Wang T, Lander ES, Sabatini DM. 2016. Large-scale single-guide RNA library construction and use for genetic screens. *Cold Spring Harb Protoc* doi: 10.1101/pdb.top086892.

Wei Y, Terns RM, Terns MP. 2015. Cas9 function and host genome sampling in type II-A CRISPR–Cas adaptation. *Genes Dev* **29:** 356–361.

Westra ER, van Erp PB, Kunne T, Wong SP, Staals RH, Seegers CL, Bollen S, Jore MM, Semenova E, Severinov K, et al. 2012. CRISPR immunity relies on the consecutive binding and degradation of negatively supercoiled invader DNA by Cascade and Cas3. *Mol Cell* **46:** 595–605.

Westra ER, Buckling A, Fineran PC. 2014. CRISPR–Cas systems: Beyond adaptive immunity. *Nat Rev Microbiol* **12:** 317–326.

Wiedenheft B, Lander GC, Zhou K, Jore MM, Brouns SJJ, van der Oost J, Doudna JA, Nogales E. 2011a. Structures of the RNA-guided surveillance complex from a bacterial immune system. *Nature* **477:** 486–489.

Wiedenheft B, van Duijn E, Bultema JB, Waghmare SP, Zhou K, Barendregt A, Westphal W, Heck AJR, Boekema EJ, Dickman MJ, et al. 2011b. RNA-guided complex from a bacterial immune system enhances target recognition through seed sequence interactions. *Proc Natl Acad Sci* **108:** 10092–10097.

Wu Y, Zhou H, Fan X, Zhang Y, Zhang M, Wang Y, Xie Z, Bai M, Yin Q, Liang D, et al. 2015. Correction of a genetic disease by CRISPR-Cas9-mediated gene editing in mouse spermatogonial stem cells. *Cell Res* **25:** 67–79.

Yin H, Xue W, Chen S, Bogorad RL, Benedetti E, Grompe M, Koteliansky V, Sharp PA, Jacks T, Anderson DG. 2014. Genome editing with Cas9 in adult mice corrects a disease mutation and phenotype. *Nat Biotechnol* **32:** 551–553.

Yosef I, Goren MG, Qimron U. 2012. Proteins and DNA elements essential for the CRISPR adaptation process in *Escherichia coli. Nucleic Acids Res* **40:** 5569–5576.

Yu Z, Ren M, Wang Z, Zhang B, Rong YS, Jiao R, Gao G. 2013. Highly efficient genome modifications mediated by CRISPR/Cas9 in *Drosophila. Genetics* **195:** 289–291.

Zegans ME, Wagner JC, Cady KC, Murphy DM, Hammond JH, O'Toole GA. 2009. Interaction between bacteriophage DMS3 and host CRISPR region inhibits group behaviors of *Pseudomonas aeruginosa. J Bacteriol* **191:** 210–219.

Zhang J, Rouillon C, Kerou M, Reeks J, Brugger K, Graham S, Reimann J, Cannone G, Liu H, Albers SV, et al. 2012. Structure and mechanism of the CMR complex for CRISPR-mediated antiviral immunity. *Mol Cell* **45:** 303–313.

Zhang Y, Heidrich N, Ampattu BJ, Gunderson CW, Seifert HS, Schoen C, Vogel J, Sontheimer EJ. 2013. Processing-independent CRISPR RNAs limit natural transformation in *Neisseria meningitidis. Mol Cell* **50:** 488–503.

Zhou Y, Zhu S, Cai C, Yuan P, Li C, Huang Y, Wei W. 2014. High-throughput screening of a CRISPR/Cas9 library for functional genomics in human cells. *Nature* **509:** 487–491.

Zuris JA, Thompson DB, Shu Y, Guilinger JP, Bessen JL, Hu JH, Maeder ML, Joung JK, Chen ZY, Liu DR. 2015. Cationic lipid-mediated delivery of proteins enables efficient protein-based genome editing in vitro and in vivo. *Nat Biotechnol* **33:** 73–80.

 Cite this introduction as *Cold Spring Harb Protoc*; doi:10.1101/pdb.top088849

CHAPTER 2

Guide RNAs: A Glimpse at the Sequences that Drive CRISPR–Cas Systems

Alexandra E. Briner[1] and Rodolphe Barrangou[1,2]

[1]*North Carolina State University, Raleigh, North Carolina 27695*

CRISPR–Cas systems provide adaptive immunity in bacteria and archaea. Although there are two main classes of CRISPR–Cas systems defined by gene content, interfering RNA biogenesis, and effector proteins, Type II systems have recently been exploited on a broad scale to develop next-generation genetic engineering and genome-editing tools. Conveniently, Type II systems are streamlined and rely on a single protein, Cas9, and a guide RNA molecule, comprised of a CRISPR RNA (crRNA) and *trans*-acting CRISPR RNA (tracrRNA), to achieve effective and programmable nucleic acid targeting and cleavage. Currently, most commercially available Cas9-based genome-editing tools use the CRISPR–Cas system from *Streptococcus pyogenes* (SpyCas9), although many orthogonal Type II systems are available for diverse and multiplexable genome engineering applications. Here, we discuss the biological significance of Type II CRISPR–Cas elements, including the tracrRNA, crRNA, Cas9, and protospacer-adjacent motif (PAM), and look at the native function of these elements to understand how they can be engineered, enhanced, and optimized for genome editing applications. Additionally, we discuss the basis for orthogonal Cas9 and guide RNA systems that would allow researchers to concurrently use multiple Cas9-based systems for different purposes. Understanding the native function of endogenous Type II CRISPR–Cas systems can lead to new Cas9 tool development to expand the genetic manipulation toolbox.

WHAT ARE CRISPR–Cas SYSTEMS?

All organisms have a need to protect themselves and develop immunity against foreign invaders. Since the mid-2000s, researchers have been studying the ability of CRISPR–Cas adaptive immune systems to protect bacteria and archaea against potentially damaging foreign nucleic acids (Mojica et al. 2005; Makarova et al. 2006; Barrangou et al. 2007; Tyson and Banfield 2008). Clustered regularly interspaced short palindromic repeats (CRISPR) and associated sequences (*cas* genes) function as the immunization records and immunity systems to provide resistance against phages, plasmids, and potentially harmful nucleic acids. Generally, CRISPR–Cas systems protect the cell in three steps: (1) *acquisition* of a nucleic acid sequence from an invader, (2) *expression* and biogenesis of small interfering RNAs, and (3) *interference* and cleavage of foreign nucleic acid of similar or homologous sequence upon re-introduction into the cell (Barrangou et al. 2007; Makarova et al. 2011, 2015; Koonin and Makarova 2013; van der Oost et al. 2014).

All CRISPR–Cas systems have repeat spacer arrays with conserved repeats between 24 and 47 nucleotides flanking variable spacer sequences that are derived from the foreign DNA of invaders (Koonin and Makarova 2013; van der Oost et al. 2014). Additionally, all three systems use the

[2]Correspondence: rbarran@ncsu.edu

Copyright © Cold Spring Harbor Laboratory Press; all rights reserved
Cite this introduction as *Cold Spring Harb Protoc*; doi:10.1101/pdb.top090902

conserved Cas1 and Cas2 proteins during the acquisition stage to detect foreign nucleic acids and store the sequences as short DNA spacers between two repeats (Arslan et al. 2014; Nuñez et al. 2014; Heler et al. 2015). The sequence stored in the repeat-spacer array is referred to as the spacer (Jansen et al. 2002), whereas the homologous sequence on the foreign DNA is referred to as the protospacer (Deveau et al. 2008). New spacers are always added to the leader end of the array (Barrangou et al. 2007); this polarized acquisition ensures that immunization events are actively transcribed and maintained for protection against the most recent invaders as transcription of the array is driven by the leader sequence which contains promoter elements (Andersson and Banfield 2008; Tyson and Banfield 2008; Wei et al. 2015). After the acquisition stage, the entire repeat-spacer array is transcribed into a single RNA transcript that contains all of the repeats and spacers encoded in the locus (Brouns et al. 2008). In order to become functional, the RNA transcript, called the precursor-CRISPR RNA (pre-crRNA), must be cleaved into smaller RNA molecules, called CRISPR RNAs (crRNAs), that contain a partial CRISPR repeat and partial spacer (Brouns et al. 2008; Deltcheva et al. 2011; Karvelis et al. 2013). Once the crRNAs have been generated, they guide the Cas effector proteins to their complementary protospacers; subsequently, the Cas proteins target, cleave, and degrade the invading complementary nucleic acid (Brouns et al. 2008; Garneau et al. 2010; Gasiunas et al. 2012). In CRISPR immunity, Cas proteins are able to distinguish target DNA from spacer-containing, nontarget self DNA by the presence of the protospacer adjacent motif (PAM) on the target (Marraffini and Sontheimer 2010). The PAM is a short nucleotide sequence flanking the protospacer on the foreign DNA (Deveau et al. 2008; Horvath et al. 2008; Mojica et al. 2009). When stored in the repeat-spacer array, a spacer sequence is not adjacent to a PAM, therefore preventing the CRISPR–Cas systems from self-targeting and cleaving the host chromosome (Deveau et al. 2008; Marraffini and Sontheimer 2010; Heler et al. 2015).

TWO CLASSES, FIVE TYPES, 16 SUBTYPES

There are two classes of CRISPR–Cas systems which can further be broken down into five system types; these different classes are distinguished by CRISPR repeat length and sequence, *cas* gene content and locus architecture, crRNA biogenesis and composition, and effector protein type and activity (Makarova et al. 2011, 2015). All CRISPR–Cas systems contain the universal *cas1* and *cas2* genes; however, these two classes can be broken down into five types based on the presence of the signature gene for each type: *cas3*, *cas9*, *cas10*, and *cpf1* for Types I, II, III, and V, respectively. Class I CRISPR–Cas systems, namely Type I and Type III systems, have been found in bacteria and archaea, and share similarities in the acquisition, expression, and interference stages of CRISPR immunity. Both of these systems use a large multiprotein complex called either Cascade (CRISPR-associated complex for antiviral defense) (Type I) or the Cmr/Csm complex (Type III) to target and cleave foreign DNA and occasionally RNA when guided by a crRNA (Brouns et al. 2008; Hale et al. 2009). Additionally, their CRISPR repeats are highly palindromic and form hairpin loops that allow a Cas ribonuclease, Cas6, to cleave the pre-crRNA into individual interfering crRNAs (Kunin et al. 2007; van der Oost et al. 2014).

Class II systems are characterized by Type II and Type V, which use a single signature protein to both bind and cleave foreign DNA (Makarova et al. 2015). Type V systems are characterized by the Cpf1 signature protein which uses a crRNA molecule containing a partial repeat and full spacer to recognize, bind, and cleave foreign sequences (Zetsche et al. 2015). Conversely, the Type II systems use the signature Cas9 protein to bind and cleave target DNA but additionally require a second RNA molecule to complex with the crRNA; additionally, RNaseIII activity is required to generate individual crRNAs in Type II systems. This second RNA molecule, called the *trans*-activating RNA (tracrRNA, pronounced "tracer-RNA"), is important in both the biogenesis of crRNAs and Cas9-guided interference against foreign DNA (Deltcheva et al. 2011; Sapranauskas et al. 2011; Gasiunas et al. 2012; Chylinski et al. 2013, 2014; Karvelis et al. 2013). The tracrRNA contains a partial complementary antirepeat at the 5′ end that allows the molecule to base pair with the repeat portion of the crRNA; this complementary region forms three structural modules within the guide RNA that are important for

Cite this introduction as *Cold Spring Harb Protoc;* doi:10.1101/pdb.top090902

Cas9 functionality: the lower stem, bulge, and upper stem (Briner et al. 2014). The 3′ end of the tracrRNA molecule comprises several hairpin structures that are key in nucleotide binding interactions with Cas9, namely the nexus and hairpins. The first hairpin structure in the tracrRNA, called the nexus, can take several structural forms, but often has a conserved nucleotide sequence in the base of the hairpin stem. Beginning with the U in the GU wobble at the base of the lower stem, the motif UnAnnC can be found in the majority of IIA nexus hairpins in tracrRNAs (Briner et al. 2014). The terminal hairpins vary in number, size, and structure, but typically contain a Rho-independent transcriptional terminator hairpin that is GC-rich and followed by a string of U's at the 3′ end. The hairpins and nexus are key factors in determining Cas9 orthogonality and cross-compatibility (Esvelt et al. 2013; Briner et al. 2014; Fonfara et al. 2014).

TYPE II CRISPR–Cas SYSTEMS

Although there are five system types to date, distribution of systems is not equal among all the groups. Overall, CRISPR–Cas systems have been detected in 47% of all bacterial and archaeal genomes. Type I systems are the most dominant by far, constituting ~60% of total CRISPR loci bacterial and archaeal genomes. Type III systems are the second most dominant system type occurring more frequently in archaea (34% of all archaeal CRISPR loci) than bacteria (25%) (Makarova et al. 2015). Type IV and V systems are definitively the rarest types of CRISPR–Cas system, constituting <2% of overall CRISPR–Cas systems. Notwithstanding the vast diversity and high rates of occurrence of CRISPR–Cas systems, Type II systems are only harbored by bacteria and are estimated to occur in <5% of all bacterial genomes (Makarova et al. 2011, 2015; Chylinski et al. 2014). Despite their rare occurrence, the system is diverse enough to be broken down into three subtypes (IIA, IIB, IIC) that have distinct Cas9 size, repeat size, array orientation, and guide RNA composition. The IIA subtype is the best characterized system and contains model systems like *Streptococcus pyogenes* (Spy) and *Streptococcus thermophilus* (Sth), and *Staphylococcus aureus* (Anders et al. 2014; Chylinski et al. 2014; Nishimasu et al. 2014; Ran et al. 2015). From these organisms, we have learned that there are two distinct groups of Cas9 sizes in the IIA subtype; these systems contain long Cas9s that are approximately 1300 amino acids in length, as characterized by the Sth-CRISPR3 locus and Spy, or contain short Cas9s around 1100 amino acids in length, as characterized by the Sth-CRISPR1 locus and *S. aureus* (Horvath et al. 2008). The IIA repeats are always 36 nucleotides in length and orientation is often easy to determine as the arrays contain degenerate repeats that differ from the consensus sequence opposite the leader end. Because the leader end of the repeat-spacer array is actively being maintained through transcription and addition of new spacers, it is hypothesized that spacer excision and repeat recombination events may lead to mutations and single nucleotide polymorphisms toward the ancestral end. The tracrRNAs are either located between the *cas9* and *cas1* genes or upstream of the *cas9* gene (Chylinski et al. 2013; Fonfara et al. 2014). Occasional reports have found antirepeats thought to be potential tracrRNAs between the *csn2* gene and the start of the repeat-spacer array and downstream from the repeat-spacer array (Chylinski et al. 2013, 2014; Briner et al. 2014).

Type IIB and IIC systems are less characterized than IIAs and have been shown to not have a clear conservation in orientation or size of repeats. Although repeats can be 36 nucleotides in length like the IIAs, some IIB repeats can be as large at 47 nucleotides. Additionally, the orientation of these repeats cannot be determined by looking for the degenerate end of the repeat-spacer array and usually must be determined through RNA-sequencing of crRNAs. The size of Cas9s in these subsystems can vary greatly. The *Neisseria meningitidis* (Nme) Cas9 (IIC) is 1082 amino acids in length and the *Legionella pneumophila* (IIB) Cas9 is 1372 amino acids in length (Esvelt et al. 2013; Hou et al. 2013; Chylinski et al. 2014). There has also not been clear characterization of the tracrRNAs in IIC and IIB systems, although often times they resemble the canonical IIA tracrRNA without the bulge module between the lower and upper stems. Interestingly, tracrRNAs for these systems still often contain a nexus hairpin like the IIA systems with a conserved sequence motif seen in all IIA tracrRNAs. Terminal hairpins are also present in IIB and IIC systems.

The tracrRNA molecule is critical for Type II CRISPR–Cas functionality and performance for both native bacterial immunity and exploitation to generate genome-editing tools. During the expression stage of native CRISPR–Cas immunity, the pre-crRNA containing all of the repeat and spacer sequences is transcribed into one molecule; the antirepeat, complementary portion of the tracrRNA is able to base pair with the repeat portions of the pre-crRNA with the aid of Cas9 (Gasiunas et al. 2012). When the double-stranded RNA (dsRNA) molecule is formed, a native RNase III, encoded by the *rnc* gene, cleaves the RNA molecule in the repeat segment, forming individual single repeat-spacer units (Deltcheva et al. 2011; Gasiunas et al. 2012). A secondary unknown RNase then trims the crRNA at the 5′ end of the spacer portion of the crRNA so that only 19–22 nucleotides of a partial CRISPR spacer is left intact. In minimal repeat-spacer arrays containing a single spacer flanked by repeats, no RNase III processing is necessary and the crRNA:tracrRNA duplex retains both flanking repeats during the interference stage (Karvelis et al. 2013).

THE tracrRNA INTERACTS CLOSELY WITH Cas9

Co-evolution of core CRISPR elements has led to divergent, orthogonal systems that are not cross-compatible (Horvath et al. 2009). Distinct evolution events have allowed the elements within a CRISPR locus, including Cas9, Cas1, CRISPR repeat, and tracrRNA, to remain compatible and adapted to one another by developing defined characteristics that distinguish the elements from one locus to be incompatible with elements from another. This close co-evolution of core elements is supported by the fact that each unique Cas9 and its tracrRNA:crRNA interact through specific binding of protein residues to guide RNA functional modules.

The tracrRNAs in Type IIA systems always contain the five canonical functional modules in addition to the spacer module: upper stem, bulge, lower stem, nexus, and hairpins (Fig. 1A; Briner et al. 2014). The first three modules, the upper stem, bulge, and lower stem, are formed by the base-pairing between the CRISPR repeat and the complementary antirepeat portion of the tracrRNA. The upper stem sequence and length are variable, but it serves as the site for dsRNA cleavage by RNaseIII during crRNA biogenesis. The bulge is variable in sequence and size but is always kinked in the same direction. The size of lower stem varies based on the system, but contains between four and eight base pairs, often ending in a G-U wobble at the base of the hairpin. Based on the Spy Cas9 guide RNA- and target DNA-bound structure, the phosphate backbone of the upper and lower stem, formed by the repeat–antirepeat binding, interacts in a sequence-independent manner with Cas9, demonstrating the crRNA:tracrRNA repeat structure is more important for Cas9 recognition than the repeat sequence (Fig. 1B; Anders et al. 2014; Nishimasu et al. 2014). The arginine bridge helix of the Cas9 binds with the base of the lower stem and the nexus, the first hairpin in the tracrRNA, restructuring into an active conformation (Jinek et al. 2012; Anders et al. 2014; Nishimasu et al. 2014). There are two key nucleotide in the nexus that directly interact with amino acid residues in SpyCas9 and drive the structural conformation in the protein that allow it to target and cleave foreign targets; one uracil in the nexus protrudes from the hairpin and interacts with an asparagine residue in the recognition lobe of Cas9, while an adenosine nucleotide binds with an arginine residue in the protein. Together, these interactions allow Cas9 to access the double-stranded DNA at the PAM site and initiate Cas9 based targeting and cleavage (Anders et al. 2014; Nishimasu et al. 2014; Sternberg et al. 2014; Heler et al. 2015; Ran et al. 2015).

BIOTECHNOLOGY APPLICATIONS

What practically allowed Cas9-based technology to catapult to the primary genetic engineering tool, arguably, was the technological advance of artificially combining the tracrRNA and crRNA through a nucleotide tetra-loop (Jinek et al. 2012; Mali et al. 2013). The artificially combined crRNA:tracrRNA duplex is referred to as a single guide RNA (sgRNA), and reduced the natural four component system

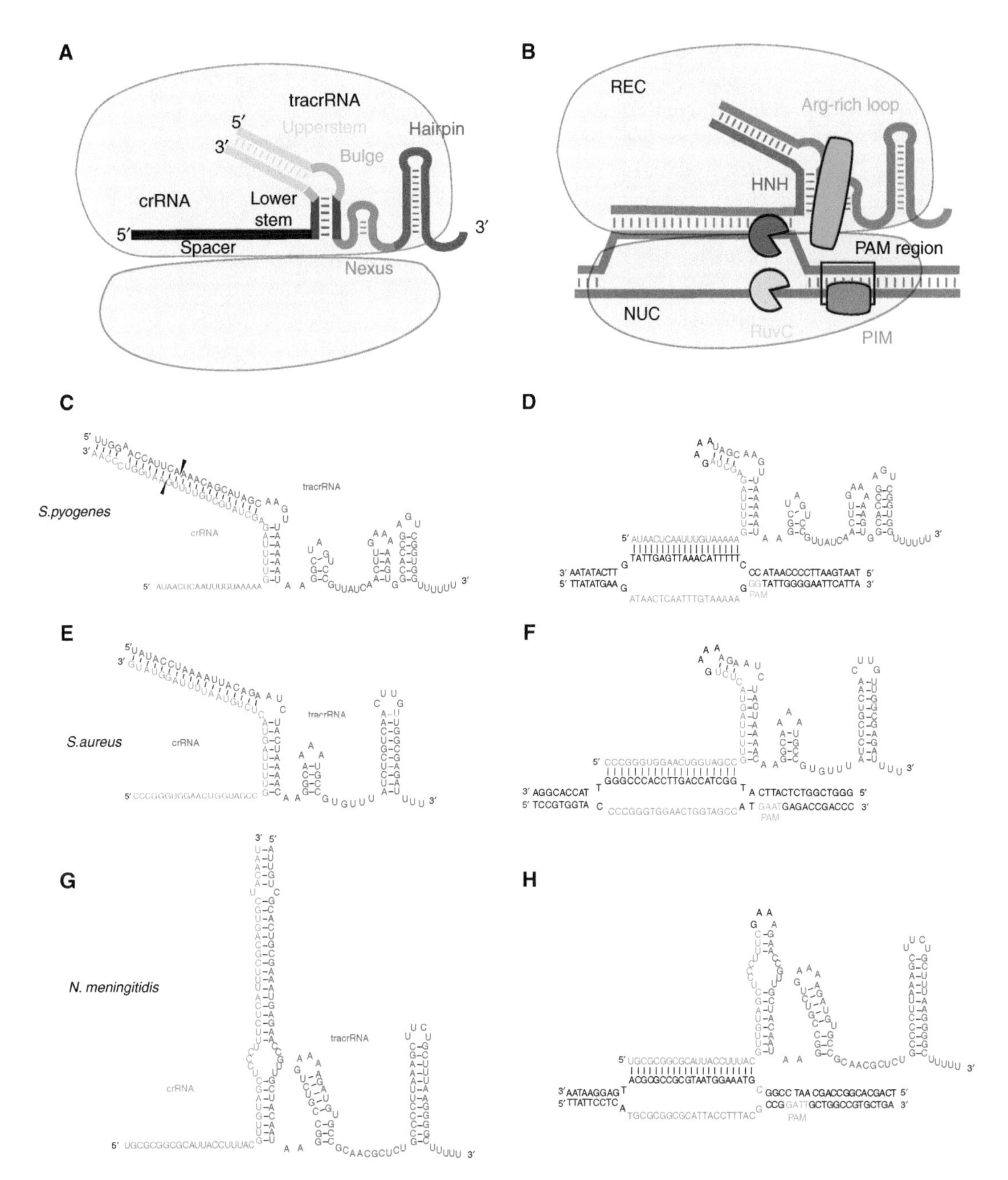

FIGURE 1. Modules of native crRNA:tracrRNA duplexes and engineered single guide RNAs interact with protein domains of Cas9. The six functional modules formed in the crRNA:tracrRNA duplex that allow it to bind Cas9 (light gray) are shown in *A*; these include the spacer (black), lower stem (dark gray), bulge (green), upper stem (yellow), nexus (blue), and terminal hairpins (red). The protein domains within Cas9 that bind and interact with the guide RNA are shown in *B*, including the arginine bridge helix (green), PAM interacting motif (blue), HNH nuclease domain (red), and RuvC nuclease domain (yellow). Additionally, the recognition (REC) lobe and nuclease (NUC) lobe of Cas9 are labeled. The full native dual crRNA (blue) and tracrRNA (red) duplex is shown for *S. pyogenes*, *S. aureus*, and *N. meningitidis* in *C*, *E*, and *G*, respectively. The single guide RNA is shown for the same organisms in complex with target DNA (black) flanked by the PAM (green), for *S. pyogenes* (Jinek et al. 2012), *S. aureus* (Ran et al. 2015), and *N. meningitidis* (Hou et al. 2013) in *D*, *F*, and *H*, respectively. The sgRNA combines the crRNA and tracrRNA by a 4-nt tetraloop (dark gray).

(Cas9:RNaseIII:crRNA:tracrRNA) to a synthetic two-component system (Cas9::sgRNA), greatly increasing ease of use and portability of the system. To exploit the DNA-targeting power of Cas9, one simply needs to understand the various components of tracrRNAs and designing corresponding guide RNAs (Fig. 1C–H).

Understanding the key modules within the tracrRNA has allowed researchers to concurrently use multiple Cas9 tools in a single cell without having to rely on the same protein, guide, and PAM during different applications. This multiplexing potential is made possible through utilization of CRISPR–Cas9 systems that contain tracrRNAs that have variable nexus and hairpin regions (Esvelt et al. 2013; Briner et al. 2014; Fonfara et al. 2014). It was determined that these two modules are the key to determine orthogonality between systems and establish the boundaries of compatibility between systems.

To date, the Cas9-based genome-editing tool box is capable of performing genetic techniques ranging from sequence-specific introduction of double-stranded breaks to gene regulation to fluorescent imaging of chromosome loci (Doudna and Charpentier 2014; Sternberg and Doudna 2015). High-throughput studies made possible by Cas9 have already begun to revolutionize the rate and depth of genome-wide surveys intended to help researchers better understand the DNA code (Cong et al. 2013; Mali et al. 2013). Although researchers are rapidly improving and expanding the specificity, range of applications, and portability of Cas9-based tools, this technology is still fairly young and presents an exciting opportunity for deeper characterization and understanding. Although great depth has already been achieved in understanding and utilization of Cas9-based genome editing tools, the surface of the potential for this technology has barely been scratched. In our accompanying protocol, we present a strategy to identify elements of bacterial immune systems that will allow us to develop next-generation Cas9-based genome editing tools (see Protocol 1: Prediction and Validation of Native and Engineered Cas9 Guide Sequences [Briner et al. 2016]). Through discovery, validation, and development of new and unique Cas9 proteins and their corresponding tracrRNAs, crRNAs, and PAMs, the CRISPR–Cas9 genome-editing technology has the potential to become as flexible as restriction enzymes and arguably as commonplace as PCR in molecular genetics laboratories.

REFERENCES

Anders C, Niewoehner O, Duerst A, Jinek M. 2014. Structural basis of PAM-dependent target DNA recognition by the Cas9 endonuclease. *Nature* **513:** 569–573.

Andersson AF, Banfield JF. 2008. Virus population dynamics and acquired virus resistance in natural microbial communities. *Science* **320:** 1047–1050.

Arslan Z, Hermanns V, Wurm R, Wagner R, Pul U. 2014. Detection and characterization of spacer integration intermediates in type I-E CRISPR-Cas system. *Nucleic Acids Res* **42:** 7884–7893.

Barrangou R, Fremaux C, Deveau H, Richards M, Boyaval P, Moineau S, Romero DA, Horvath P. 2007. CRISPR provides acquired resistance against viruses in prokaryotes. *Science* **315:** 1709–1712.

Briner AE, Donohoue PD, Gomaa AA, Selle K, Slorach EM, Nye CH, Haurwitz RE, Beisel CL, May AP, Barrangou R. 2014. Guide RNA functional modules direct Cas9 activity and orthogonality. *Mol Cell* **56:** 333–339.

Briner AE, Henriksen ED, Barrangou R. 2016. Prediction and validation of native and engineered Cas9 guide sequences. *Cold Spring Harb Protoc* doi: 10.1101/pdb.prot086785.

Brouns SJ, Jore MM, Lundgren M, Westra ER, Slijkhuis RJ, Snijders AP, Dickman MJ, Makarova KS, Koonin EV, van der Oost J. 2008. Small CRISPR RNAs guide antiviral defense in prokaryotes. *Science* **321:** 960–964.

Chylinski K, Le Rhun A, Charpentier E. 2013. The tracrRNA and Cas9 families of type II CRISPR-Cas immunity systems. *RNA Biol* **10:** 726–737.

Chylinski K, Makarova KS, Charpentier E, Koonin EV. 2014. Classification and evolution of type II CRISPR-Cas systems. *Nucleic Acids Res* **42:** 6091–6105.

Cong L, Ran FA, Cox D, Lin S, Barretto R, Habib N, Hsu PD, Wu X, Jiang W, Marraffini LA, et al. 2013. Multiplex genome engineering using CRISPR/Cas systems. *Science* **339:** 819–823.

Deltcheva E, Chylinski K, Sharma CM, Gonzales K, Chao Y, Pirzada ZA, Eckert MR, Vogel J, Charpentier E. 2011. CRISPR RNA maturation by trans-encoded small RNA and host factor RNase III. *Nature* **471:** 602–607.

Deveau H, Barrangou R, Garneau JE, Labonté J, Fremaux C, Boyaval P, Romero DA, Horvath P, Moineau S. 2008. Phage response to CRISPR-encoded resistance in *Streptococcus thermophilus*. *J Bacteriol* **190:** 1390–1400.

Doudna JA, Charpentier E. 2014. Genome editing. The new frontier of genome engineering with CRISPR-Cas9. *Science* **346:** 1258096.

Esvelt KM, Mali P, Braff JL, Moosburner M, Yaung SJ, Church GM. 2013. Orthogonal Cas9 proteins for RNA-guided gene regulation and editing. *Nat Methods* **10:** 1116–1121.

Fonfara I, Le Rhun A, Chylinski K, Makarova KS, Lécrivain A-L, Bzdrenga J, Koonin EV, Charpentier E. 2014. Phylogeny of Cas9 determines functional exchangeability of dual-RNA and Cas9 among orthologous type II CRISPR-Cas systems. *Nucleic Acids Res* **42:** 2577–2590.

Garneau JE, Dupuis ME, Villion M, Romero DA, Barrangou R, Boyaval P, Fremaux C, Horvath P, Magadan AH, Moineau S. 2010. The CRISPR/Cas bacterial immune system cleaves bacteriophage and plasmid DNA. *Nature* **468:** 67–71.

Gasiunas G, Barrangou R, Horvath P, Siksnys V. 2012. Cas9-crRNA ribonucleoprotein complex mediates specific DNA cleavage for adaptive immunity in bacteria. *Proc Natl Acad Sci* **109:** E2579–E2586.

Hale CR, Zhao P, Olson S, Duff MO, Graveley BR, Wells L, Terns RM, Terns MP. 2009. RNA-guided RNA cleavage by a CRISPR RNA-Cas protein complex. *Cell* **139:** 945–956.

Heler R, Samai P, Modell JW, Weiner C, Goldberg GW, Bikard D, Marraffini LA. 2015. Cas9 specifies functional viral targets during CRISPR-Cas adaptation. *Nature* **519:** 199–202.

Horvath P, Romero DA, Coute-Monvoisin AC, Richards M, Deveau H, Moineau S, Boyaval P, Fremaux C, Barrangou R. 2008. Diversity, activity, and evolution of CRISPR loci in *Streptococcus thermophilus*. *J Bacteriol* **190:** 1401–1412.

Horvath P, Coute-Monvoisin AC, Romero DA, Boyaval P, Fremaux C, Barrangou R. 2009. Comparative analysis of CRISPR loci in lactic acid bacteria genomes. *Int J Food Microbiol* **131:** 62–70.

Hou Z, Zhang Y, Propson NE, Howden SE, Chu LF, Sontheimer EJ, Thomson JA. 2013. Efficient genome engineering in human pluripotent

 Cite this introduction as *Cold Spring Harb Protoc*; doi:10.1101/pdb.top090902

stem cells using Cas9 from *Neisseria meningitidis. Proc Natl Acad Sci* **110:** 15644–15649.

Jansen R, Embden JD, Gaastra W, Schouls LM. 2002. Identification of genes that are associated with DNA repeats in prokaryotes. *Mol Microbiol* **43:** 1565–1575.

Jinek M, Chylinski K, Fonfara I, Hauer M, Doudna JA, Charpentier E. 2012. A programmable dual-RNA–guided DNA endonuclease in adaptive bacterial immunity. *Science* **337:** 816–821.

Karvelis T, Gasiunas G, Miksys A, Barrangou R, Horvath P, Siksnys V. 2013. crRNA and tracrRNA guide Cas9-mediated DNA interference in *Streptococcus thermophilus. RNA Biol* **10:** 841–851.

Koonin EV, Makarova KS. 2013. CRISPR-Cas: Evolution of an RNA-based adaptive immunity system in prokaryotes. *RNA Biol* **10:** 679–686.

Kunin V, Sorek R, Hugenholtz P. 2007. Evolutionary conservation of sequence and secondary structures in CRISPR repeats. *Genome Biol* **8:** R61.

Makarova KS, Grishin NV, Shabalina SA, Wolf YI, Koonin EV. 2006. A putative RNA-interference-based immune system in prokaryotes: Computational analysis of the predicted enzymatic machinery, functional analogies with eukaryotic RNAi, and hypothetical mechanisms of action. *Biol Direct* **1:** 7.

Makarova KS, Haft DH, Barrangou R, Brouns SJ, Charpentier E, Horvath P, Moineau S, Mojica FJ, Wolf YI, Yakunin AF. 2011. Evolution and classification of the CRISPR–Cas systems. *Nat Rev Microbiol* **9:** 467–477.

Makarova KS, Wolf YI, Alkhnbashi OS, Costa F, Shah SA, Saunders SJ, Barrangou R, Brouns SJJ, Charpentier E, Haft DH, et al. 2015. An updated evolutionary classification of CRISPR-Cas systems. *Nat Rev Microbiol* **13:** 722–736.

Mali P, Yang L, Esvelt KM, Aach J, Guell M, DiCarlo JE, Norville JE, Church GM. 2013. RNA-guided human genome engineering via Cas9. *Science* **339:** 823–826.

Marraffini LA, Sontheimer EJ. 2010. Self versus non-self discrimination during CRISPR RNA-directed immunity. *Nature* **463:** 568–571.

Mojica FJ, Diez-Villasenor C, Garcia-Martinez J, Soria E. 2005. Intervening sequences of regularly spaced prokaryotic repeats derive from foreign genetic elements. *J Mol Evol* **60:** 174–182.

Mojica FJ, Diez-Villasenor C, Garcia-Martinez J, Almendros C. 2009. Short motif sequences determine the targets of the prokaryotic CRISPR defence system. *Microbiol* **155:** 733–740.

Nishimasu H, Ran FA, Hsu PD, Konermann S, Shehata SI, Dohmae N, Ishitani R, Zhang F, Nureki O. 2014. Crystal structure of Cas9 in complex with guide RNA and target DNA. *Cell* **156:** 935–949.

Nuñez JK, Kranzusch PJ, Noeske J, Wright AV, Davies CW, Doudna JA. 2014. Cas1–Cas2 complex formation mediates spacer acquisition during CRISPR–Cas adaptive immunity. *Nat Struct Mol Biol* **21:** 528–534.

Ran FA, Cong L, Yan WX, Scott DA, Gootenberg JS, Kriz AJ, Zetsche B, Shalem O, Wu X, Makarova KS, et al. 2015. In vivo genome editing using *Staphylococcus aureus* Cas9. *Nature* **520:** 186–191.

Sapranauskas R, Gasiunas G, Fremaux C, Barrangou R, Horvath P, Siksnys V. 2011. The *Streptococcus thermophilus* CRISPR/Cas system provides immunity in *Escherichia coli. Nucleic Acids Res* **39:** 9275–9282.

Sternberg SH, Doudna JA. 2015. Expanding the Biologist's Toolkit with CRISPR-Cas9. *Mol Cell* **58:** 568–574.

Sternberg SH, Redding S, Jinek M, Greene EC, Doudna JA. 2014. DNA interrogation by the CRISPR RNA-guided endonuclease Cas9. *Nature* **507:** 62–67.

Tyson GW, Banfield JF. 2008. Rapidly evolving CRISPRs implicated in acquired resistance of microorganisms to viruses. *Environ Microbiol* **10:** 200–207.

van der Oost J, Westra ER, Jackson RN, Wiedenheft B. 2014. Unravelling the structural and mechanistic basis of CRISPR-Cas systems. *Nat Rev Microbiol* **12:** 479–492.

Wei Y, Chesne MT, Terns RM, Terns MP. 2015. Sequences spanning the leader-repeat junction mediate CRISPR adaptation to phage in *Streptococcus thermophilus. Nucleic Acids Res* **43:** 1749–1758.

Zetsche B, Gootenberg JS, Abudayyeh OO, Slaymaker IM, Makarova KS, Essletzbichler P, Volz SE, Joung J, van der Oost J, Regev A, et al. 2015. Cpf1 is a single RNA-guided endonuclease of a class 2 CRISPR-Cas system. *Cell* **163:** 759–771.

Protocol 1

Prediction and Validation of Native and Engineered Cas9 Guide Sequences

Alexandra E. Briner,[1] Emily D. Henriksen,[1] and Rodolphe Barrangou[1,2]

[1]*Department of Food, Bioprocessing and Nutrition Sciences, North Carolina State University, Raleigh, North Carolina 27695*

Cas9-based technologies rely on native elements of Type II CRISPR–Cas bacterial immune systems, including the *trans*-activating CRISPR RNA (tracrRNA), CRISPR RNA (crRNA), Cas9 protein, and protospacer-adjacent motif (PAM). The tracrRNA and crRNA form an RNA duplex that guides the Cas9 endonuclease to complementary nucleic acid sequences. Mechanistically, Cas9 initiates interactions by binding to the target PAM sequence and interrogating the target DNA in a 3′-to-5′ manner. Complementarity between the guide RNA and the target DNA is key. In natural systems, precise cleavage occurs when the target DNA sequence contains a PAM flanking a sequence homologous to the crRNA spacer sequence. Currently, the majority of commercial Cas9-based genome-editing tools are derived from the Type II CRISPR–Cas system of *Streptococcus pyogenes*. However, a diverse set of Type II CRISPR–Cas systems exist in nature that are potentially valuable for genome engineering applications. Exploitation of these systems requires prediction and validation of both native and engineered dual and single guide RNAs to drive Cas9 functionality. Here, we discuss how to identify the elements of these immune systems to develop next-generation Cas9-based genome-editing tools. We first discuss how to predict tracrRNA sequences and suggest a method for designing single guide RNAs containing only critical structural modules. We then outline how to predict the PAM sequence, which is crucial for determining potential targets for Cas9. Finally, validation of the system elements through transcriptome analysis and interference assays is essential for developing next-generation Cas9-based genome-editing tools.

MATERIALS

It is essential that you consult the appropriate Material Safety Data Sheets and your institution's Environmental Health and Safety Office for proper handling of equipment and hazardous materials used in this protocol.

Reagents

Direct-zol RNA MiniPrep Kit (Zymo Research)

The Direct-zol RNA MiniPrep Kit from Zymo Research is highly recommended for extraction of RNA as it allows purification of RNA molecules as small as 17 nt, thereby retaining the smaller crRNAs that many traditional kits discard.

Next-generation sequencing reagents (HiSeq or MiSeq; Illumina)
Sample for RNA extraction
TruSeq Small RNA Sample Preparation Kit (Illumina)

[2]Correspondence: rbarran@ncsu.edu

Copyright © Cold Spring Harbor Laboratory Press; all rights reserved
Cite this protocol as *Cold Spring Harb Protoc*; doi:10.1101/pdb.prot086785

Equipment

Basic Local Alignment Search Tool (BLAST) and Conserved Domain Database (CCD) (National Center for Biotechnology Information [NCBI])

CRISPR identification program

CRISPRfinder (http://crispr.u-psud.fr/Server/) is a web-based CRISPR identification platform that allows users to upload their own sequences of interest as well as browse genomes annotated directly from the NCBI database (Grissa et al. 2007). CRISPR Recognition Tool (CRT) (http://www.room220.com/crt/ [Bland et al. 2007]) is available as a command-line tool or plug-in for several commercially available graphical user interface (GUI) bioinformatics platforms (e.g., Geneious9 [Biomatters]). Both programs identify potential CRISPR spacers and repeats, based on a repeat-finding algorithm.

Genome sequence of interest

Quality control software for RNA sequencing data (e.g., FastQC)

RNA folding prediction software (e.g., NUPACK; http://www.nupack.org/ [Zadeh et al. 2011])

Sequence motif identification program (e.g., WebLogo [Crooks et al. 2004])

Short read sequence alignment tool (e.g., Bowtie 2 [Langmead and Salzberg 2012])

METHOD

Predicting CRISPR Repeats and *cas* Genes In Silico

Type II CRISPR–Cas systems have been identified in only 5% of bacterial genomes and are yet to be detected in archaeal genomes (Makarova et al. 2011; Chylinski et al. 2013). These are the least widely distributed CRISPR–Cas systems overall and are enriched in Firmicutes and Actinobacteria. Notably, lactic acid bacteria and bacteria closely related with humans have been shown to contain Type II systems more frequently.

1. Upload the genome sequence of interest to a CRISPR identification program.
2. Identify the CRISPR repeat consensus sequence (Fig. 1A).

   Although documented CRISPR repeats have been between 24 and 47 nt in length, overall most CRISPR repeats are between 29 and 36 nt. For Type II systems in particular, CRISPR repeats are generally 36 nt in length (Makarova et al. 2011; Chylinski et al. 2013).
3. Search the region flanking the repeat-spacer array for *cas* genes, looking specifically for the universal *cas1* and the signature Type II gene, *cas9*, to ensure you are investigating a Type II CRISPR–Cas system (Fig. 1B). Confirm the *cas* gene annotation using BLAST or CDD.

   For the Type II subtypes, the prototypical Cas9 proteins are from the CRISPR–Cas systems found in Streptococcus thermophilus *CNRZ 1066,* Legionella pneumophila *str.* Paris, *and* Neisseria lactamica *020-06, representing the II-A, II-B, and II-C subtypes, respectively (Makarova et al. 2011). Cas9 should contain the two nickase domains RuvC and HNH (Sapranauskas et al. 2011).*
4. Determine the orientation of the array by identifying the leader sequence and the terminal repeat (Fig. 1C).

   The leader sequence is adjacent to the first repeat and initiates transcription of the repeat-spacer array. It is often AT-rich. The terminal end of the array can be determined by identifying CRISPR repeats with single point mutations. The terminal repeat often differs from the consensus repeat sequence by several mutations, frequently at the 3′ end. The correct orientation of the array should start at the leader on the 5′ side and end with the terminal repeat on the 3′ side.

Predicting tracrRNA In Silico

5. To search for the antirepeat portion of the tracrRNA, perform a local BLAST nucleotide alignment (Altschul et al. 1997) between the consensus CRISPR repeat sequence and DNA sequences within one of four noncoding regions: within 500 nt upstream of Cas9, between *cas9* and *cas1*, between *csn2/cas4* and the repeat-spacer array, or within 500 nt downstream from the repeat-spacer array (Chylinkski et al. 2014) (Fig. 1D). Use the following BLAST parameters.

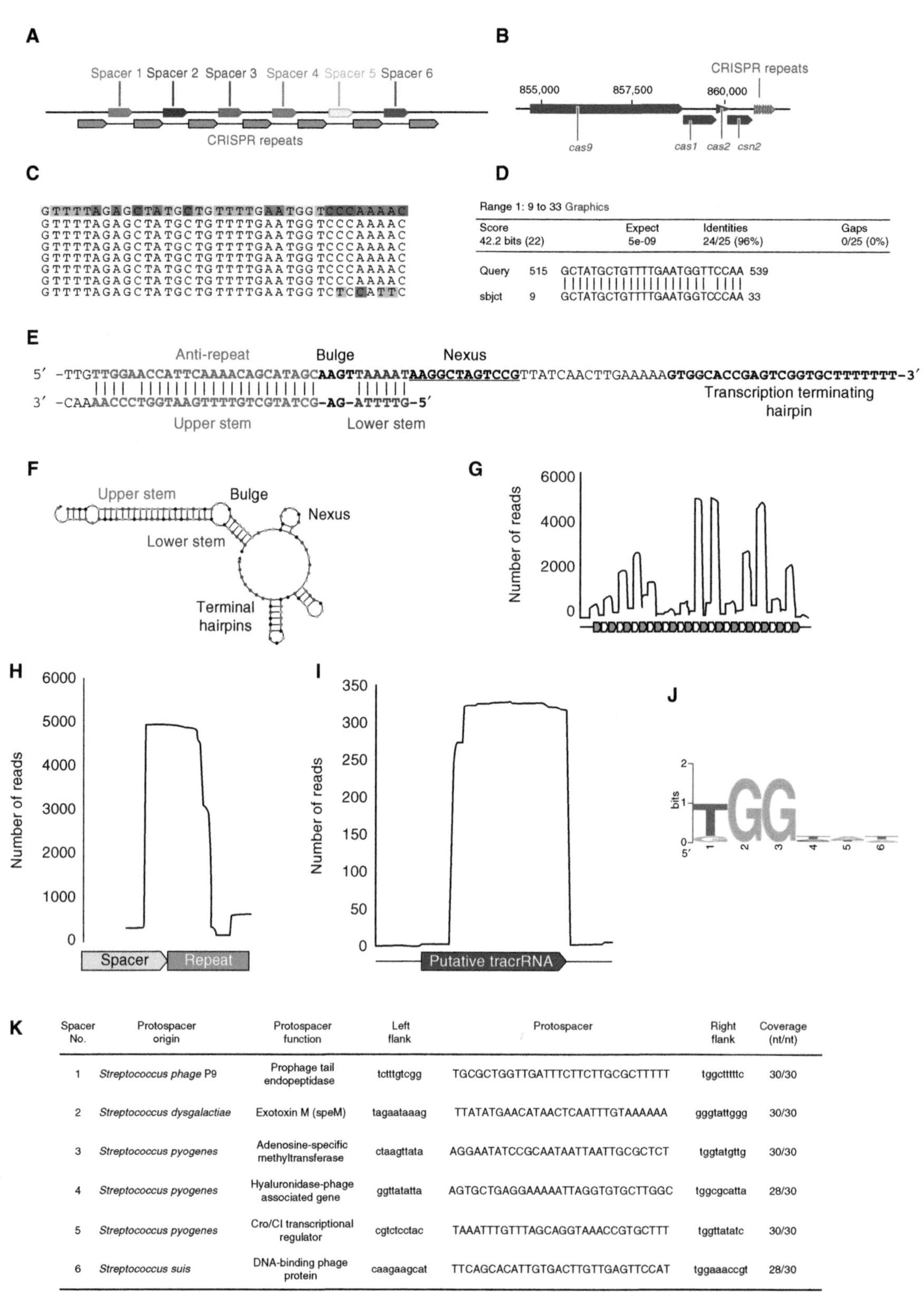

Spacer No.	Protospacer origin	Protospacer function	Left flank	Protospacer	Right flank	Coverage (nt/nt)
1	*Streptococcus phage* P9	Prophage tail endopeptidase	tctttgtcgg	TGCGCTGGTTGATTTCTTCTTGCGCTTTTT	tggctttttc	30/30
2	*Streptococcus dysgalactiae*	Exotoxin M (speM)	tagaataaag	TTATATGAACATAACTCAATTTGTAAAAAA	gggtattggg	30/30
3	*Streptococcus pyogenes*	Adenosine-specific methyltransferase	ctaagttata	AGGAATATCCGCAATAATTAATTGCGCTCT	tggtatgttg	30/30
4	*Streptococcus pyogenes*	Hyaluronidase-phage associated gene	ggttatatta	AGTGCTGAGGAAAAATTAGGTGTGCTTGGC	tggcgcatta	28/30
5	*Streptococcus pyogenes*	Cro/CI transcriptional regulator	cgtctcctac	TAAATTTGTTTAGCAGGTAAACCGTGCTTT	tggttatatc	30/30
6	*Streptococcus suis*	DNA-binding phage protein	caagaagcat	TTCAGCACATTGTGACTTGTTGAGTTCCAT	tggaaaccgt	28/30

FIGURE 1. Annotation and validation of Type II CRISPR–Cas9 system elements. (*A*) Visualization output (in the Geneious7 graphical user interface) of the CRISPR Recognition Tool (CRT) (Bland et al. 2007), which is used to identify CRISPR repeats and spacers. Variable spacers (colored arrows) are flanked by conserved CRISPR repeats (green arrows). (*B*) Identification and confirmation of *cas* genes adjacent to the CRISPR repeat-spacer array. Predicted *cas* genes should be confirmed using NCBI's BLAST or CDD to ensure correct annotation of the locus. (*C*) The terminal repeat often contains nucleotide mutations that can be used to determine the directionality of the CRISPR array. (*Legend continues on following page.*)

Algorithm	Somewhat similar sequences (blastn)
Word size	7
Match/mismatch scores	1, –2
Gap costs	Existence: 1, Extension: 2

See Troubleshooting.

6. From the alignment results, identify "potential tracrRNA" sequence(s) (Fig. 1E).

   The alignment should cover between 8 nt (usually for Type II-B and II-C systems) and three-quarters of the length of the CRISPR repeat (~36 nt). For example, if you have a 36-nt repeat, look for an antirepeat in which the alignment spans 8–27 nt of the repeat sequence.

7. In the genome, locate the antirepeat of the tracrRNA (Fig. 1E). Extend the tracrRNA search in the 3′ direction until you find a sequence that resembles a Rho-independent transcription terminator (i.e., a GC-rich hairpin followed by a string of Ts). If there is not an obvious transcriptional terminator, extend the tracrRNA search in the 5′ direction, looking for a similar structure flanked by an A-rich string. If the tracrRNA is encoded in the 3′- to-5′ direction, use the reverse complement of the sequence during further analyses.

   Typically, tracrRNA sequences are at least 50 nt and <150 nt (Chylinski et al. 2013; Briner et al. 2014). Predicting directionality of the antirepeat and tracrRNA can be difficult. The directionality of the CRISPR repeat sequence can help determine the orientation of the tracrRNA. The 5′ end of the CRISPR repeat generally starts at a G that forms a G-U wobble with the tracrRNA. Additionally, the CRISPR repeat will have ~5–7 nt that base-pair with the tracrRNA to form the lower stem, followed by an unpaired segment on the tracrRNA that forms the bulge. When looking for antirepeat sequences, only the upper stem is usually identified because of the long segment of complementary base-pairing between the crRNA repeat and the tracrRNA antirepeat followed by nonpairing nucleotides in the bulge (Briner et al. 2014).

8. Using an RNA folding prediction software like NUPACK, predict the secondary structures of the crRNA and tracrRNA duplex (Fig. 1F). Use a folding algorithm that will allow for G-U base

---

FIGURE 1. (*Continued*) The consensus CRISPR repeat sequence is shown on the top line highlighted in colored boxes. The final CRISPR repeat has four mutated nucleotides at the 3′ end (highlighted); thus, it is likely to be the terminal repeat. (*D*) The alignment output from BLAST local alignment identifies the antirepeat portion of the tracrRNA that forms the upper stem (Briner et al. 2014). The antirepeat covers 1/5 to 2/3 of the entire repeat sequence. The "Query" sequence is the genomic region upstream of the *cas9* gene in *Streptococcus pyogenes*. The "sbjct" (Subject) sequence is the consensus CRISPR repeat sequence from *Streptococcus pyogenes*. The antirepeat detected by the local alignment had 96% identity to a 25-nt stretch of the CRISPR repeat with zero gaps. (*E*) The tracrRNA in the CRISPR locus is identified by locating the antirepeat segment from the local alignment. The antirepeat forms the upper stem (red) portion of the crRNA:tracrRNA duplex. Extending through the CRISPR repeat, there are several unpaired nucleotides that form the bulge (bold), followed by reestablished complementarity to form the lower stem module (blue). Adjacent to the lower stem is the nexus module (underlined) that forms a small hairpin structure in the RNA secondary structure prediction (Briner et al. 2014). (*F*) The predicted tracrRNA search is extended through a Rho-independent transcriptional terminator, such as the final bolded nucleotides in the *Streptococcus pyogenes* tracrRNA sequence. The RNA structure prediction from NUPACK (Zadeh et al. 2011) contains the five functional modules formed when the crRNA and tracrRNA form a duplex molecule (Briner et al. 2014). (*G*) RNA sequencing reads that map to the CRISPR repeat-spacer array can be used to determine the processing boundaries for the crRNAs. The conserved repeats (green arrows) flank variable spacer sequences (gray arrows). The crRNAs can be some of the most highly transcribed small RNAs in the cell. (*H*) The 5′ boundary of the crRNA will be in the spacer sequence as a result of cellular nuclease activity. The 3′ boundary of the crRNA in the CRISPR repeat is matured by RNase III processing activity when the pre-crRNA is complexed with the tracrRNA (Deltcheva et al. 2011; Karvelis et al. 2013). (*I*) The predicted tracrRNA can be confirmed through RNA sequencing analyses. The 5′ end of the tracrRNA is matured by RNase III activity when the tracrRNA is interacting with the pre-crRNA. The 3′ end of tracrRNA either is the transcript terminator or is processed by cellular nucleases. The predicted tracrRNA is often longer than the RNA sequencing-confirmed tracrRNA, as the predicted sequence contains a portion that is removed during crRNA biogenesis (Deltcheva et al. 2011; Karvelis et al. 2013). (*J*) A motif detection program, like WebLogo (Crooks et al. 2004), is used to identify the conserved PAM in the region downstream from the protospacer. The height of each nucleotide correlates to its conservation at each position. (*K*) The table of protospacer hits shows that the spacer sequences match phage and streptococci sequences in publicly available data. The best matches (>90% identity over the entire spacer sequence) can be used to extract flanking regions and predict the PAM.

wobbles. For NUPACK, ensure that tracrRNA and crRNA sequences are entered in the 5′ to 3′ direction. Use the following options (required).

Nucleic acid type	RNA
Number of strand species	2
Maximum complex size	2
Concentration of strand1	1 µM
Concentration of strand2	1 µM

See Troubleshooting.

9. Identify the secondary structures established by Briner et al. (2014), including the upper stem, bulge, lower stem, nexus, and hairpins (Fig. 1F).

   See Troubleshooting.

Confirming crRNA and tracrRNA Boundaries

After in silico determination of the putative crRNA and tracrRNA boundaries, validation of sequences by RNA sequencing is strongly recommended. Steps 12 and 13 can be performed by various programs available through open source or commercial platforms. A program such as FastQC is recommended to assess the quality and adapter content of the reads both before and after processing.

10. Extract RNA using the Direct-zol RNA MiniPrep Kit.
11. Size-select for small RNAs (between 17 and 200 nt) and perform deep sequencing using next-generation sequencing.

    We recommend using the TruSeq Small RNA Sample Preparation Kit followed by HiSeq or MiSeq Illumina sequencing with single-end 150-nt read lengths.

12. After demultiplexing the samples, trim and filter the sequencing reads to remove adapters and poor quality bases. First, remove adapters specific to the type of sequencing performed. Next, trim to remove poor quality bases using an error probability limit of at least 0.01 (Phred 20).

    More stringent trimming to 0.001 (Phred 30) is recommended.

13. After trimming, filter reads to remove sequences shorter than 15 nt, as they can map indiscriminately to the reference sequence.
14. Using a short read alignment algorithm like Bowtie 2, map the trimmed and filtered reads to the reference genome.

    A coverage map of the CRISPR–Cas locus (or individual components thereof) can be used to determine the boundaries of the crRNAs and tracrRNA as depicted in Figure 1H,I.

Designing Single Guide RNA from Chimeric crRNA:tracrRNA

15. Create a chimeric, single guide RNA by linking the 3′ end of the crRNA to the 5′ end of the tracrRNA sequences in the upper stem portion of the crRNA:tracrRNA duplex using an artificial nucleotide tetraloop composed of noncomplementary nucleotides (Jinek et al. 2012).
    - If you confirmed the boundaries of the tracrRNA and crRNAs using RNA sequencing, use the RNase III processing as the artificial linker point (Deltcheva et al. 2011).
    - If you did not confirm the boundaries, join the two molecules between 2 and 6 nt above the bulge in the upper stem.

Predicting PAM Sequences In Silico

16. Identify the protospacer sequence that each spacer was derived from by extracting the spacer sequences from the CRISPR array and searching for homologous sequence in publically available databases (NCBI) (Fig. 1K). Use only protospacers that show 90% identity over the entire spacer length for further analyses. Look for hits in plasmids, phages, and prophage regions of the chromosome.

 Cite this protocol as *Cold Spring Harb Protoc*; doi:10.1101/pdb.prot086785

If using BLASTn to identify protospacers, the following databases are recommended:

- *Nucleotide collection (nr/nt)*
- *Whole-genome shotgun contigs (wgs)*
- *Organism: Enter the genus of your organism (*Streptococcus *[taxid:1301]).*
- *WGS Project: Select metagenomes that would contain your organism.*

When looking for protospacers, manual curation of BLAST results is the key to finding high-quality matches. Ensure that perfect matches are to actual protospacers and are not, in fact, matches to identical spacer sequences in other strains of your select species (Deveau et al. 2008; Mojica et al. 2009; Shah et al. 2013).

See Troubleshooting.

17. Extract 10 nucleotide-flanking regions from both edges of the protospacer. If the BLAST result did not cover the entire spacer sequence, extend the protospacer region to cover the entire spacer length and then extract the flanking regions.

    Typically, PAMs flank the 3′ end of the protospacers in Type II systems.

18. Using a motif-identifying program like WebLogo, identify the conservation of nucleotides at each position within the flanking regions (Fig. 1J). If using WebLogo, under the Advanced Logo Options, select DNA/RNA for Sequence Type.

    Ensure that you have the correct directionality of the spacer and protospacer sequences.

TROUBLESHOOTING

Problem (Step 5): An antirepeat cannot be found in any of the regions searched.

Solution: First, try broadening your search window. Search within the flanking 1000 nt and within the *cas* genes. Additionally, try using a less stringent match/mismatch and gap cost matrix that will not penalize mismatches as harshly. Finally, if the genome is in draft status, the rest of the CRISPR locus and tracrRNA may be on a separate contig. Search the rest of the genome for additional parts of the repeat-spacer array that did not assemble well, and search the flanking regions for a tracrRNA.

Problem (Steps 8 and 9): The RNA folding prediction does not form a crRNA:tracrRNA duplex or does not contain the five modules established by Briner et al. (2014) and one RNA strand forms secondary structures with itself.

Solution: Ensure that both sequences are entered in the 5′-to-3′ direction. If the program does not predict any strand1–strand2 interactions, try entering the reverse complement of the crRNA sequence. If the folding prediction does not contain the modules established by Briner et al. (2014), try adjusting the length of the crRNA and tracrRNA sequences to decrease the amount of self-binding from the RNA strands. If this still does not yield a crRNA:tracrRNA duplex, you may not have the correct directionality of the tracrRNA. Try extending the tracrRNA search to the opposite side of the identified antirepeat (i.e., if you extended the tracrRNA on the 3′ side but did not form a crRNA:tracrRNA complex, extend the tracrRNA on the 5′ side).

Problem (Step 16): There were not enough positive, high-quality protospacer hits to infer the PAM sequence.

Solution: Oftentimes, the same CRISPR–Cas system can be found in other strains of the same organism. Identify other strains that contain an identical system (with identical repeats and highly similar Cas proteins as determined through an alignment). Use the spacer sequences from these strains to complement your PAM search.

DISCUSSION

The activity of predicted Type II CRISPR–Cas system elements should be validated through further analyses. In vivo interference testing of native systems in their bacterial backgrounds demonstrates the system's native ability to target foreign DNA. Additionally, biochemical in vitro testing of Cas9 and guide sequences can help confirm dual nickase activity (for genesis of double-stranded breaks) and possibly determine efficiency. Finally, editing with engineered CRISPR–Cas systems in vivo can help determine efficiency and specificity (Cong et al. 2013; Mali et al. 2013; Doudna and Charpentier 2014; Hsu et al. 2014; Sander and Joung 2014). Although off-target effects have been exaggerated, they must be somewhat quantified before in vivo implementation. Regarding efficiency, several items should be considered when designing guide RNAs and selecting targets, including avoiding PAM redundancies in the spacer sequence (especially the seed sequence), avoiding homopolymeric runs in target sequences, avoiding hairpin-forming sequences within the spacer, and ensuring the spacer sequence does not compromise the overall guide structure. For concurrent use of various Cas9 proteins and their corresponding guides, it is critical to ensure that these systems are orthogonal and do not cross-react (Cong et al. 2013; Esvelt et al. 2013; Fonfara et al. 2014). Accordingly, users should ensure that various guides with incompatible structures, nexus sequences, and PAMs are selected appropriately (Briner et al. 2014). We anticipate the development of novel CRISPR–Cas9 systems will expand the toolbox and open new avenues for diverse genetic engineering applications including genome editing, transcriptional control, imaging, epigenetics, and remodeling of chromosomes.

REFERENCES

Altschul SF, Madden TL, Schäffer AA, Zhang J, Zhang Z, Miller W, Lipman DJ. 1997. Gapped BLAST and PSI-BLAST: A new generation of protein database search programs. *Nucleic Acids Res* **25:** 3389–3402.

Bland C, Ramsey TL, Sabree F, Lowe M, Brown K, Kyrpides NC, Hugenholtz P. 2007. CRISPR Recognition Tool (CRT): A tool for automatic detection of clustered regularly interspaced palindromic repeats. *BMC Bioinformatics* **8:** 209.

Briner AE, Donohoue PD, Gomaa AA, Selle K, Slorach EM, Nye CH, Haurwitz RE, Beisel CL, May AP, Barrangou R. 2014. Guide RNA functional modules direct Cas9 activity and orthogonality. *Mol Cell* **53:** 333–339.

Chylinski K, Le Rhun A, Charpentier E. 2013. The tracrRNA and Cas9 families of type II CRISPR–Cas immunity systems. *RNA Biol* **10:** 726–737.

Chylinkski K, Makarova KS, Charpentier E, Koonin EV. 2014. Classification and evolution of type II CRISPR-Cas systems. *Nucleic Acids Res* **42:** 6091–6105.

Cong L, Ran FA, Cox D, Lin S, Barretto R, Habib N, Hsu PD, Wu X, Jiang W, Marraffini LA, et al. 2013. Multiplex genome engineering using CRISPR/Cas systems. *Science* **339:** 819–823.

Crooks GE, Hon G, Chandonia JM, Brenner SE. 2004. WebLogo: A sequence logo generator. *Genome Res* **14:** 1188–1190.

Deltcheva E, Chylinski K, Sharma CM, Gonzales K, Chao Y, Pirzada ZA, Eckert MR, Vogel J, Charpentier E. 2011. CRISPR RNA maturation by *trans*-encoded small RNA and host factor RNase III. *Nature* **471:** 602–607.

Deveau H, Barrangou R, Garneau JE, Labonté J, Fremaux C, Boyaval P, Romero DA, Horvath P, Moineau S. 2008. Phage response to CRISPR-encoded resistance in *Streptococcus thermophilus*. *J Bacteriol* **190:** 1390–1400.

Doudna JA, Charpentier E. 2014. Genome editing. The new frontier of genome engineering with CRISPR–Cas9. *Science* **346:** 1258096.

Esvelt KM, Mali P, Braff JL, Moosburner M, Yaung SJ, Church GM. 2013. Orthogonal Cas9 proteins for RNA-guided gene regulation and editing. *Nat Methods* **10:** 1116–1121.

Fonfara I, Le Rhun A, Chylinski K, Makarova KS, Lécrivain AL, Bzdrenga J, Koonin EV, Charpentier E. 2014. Phylogeny of Cas9 determines functional exchangeability of dual-RNA and Cas9 among orthologous type II CRISPR–Cas systems. *Nucleic Acids Res* **42:** 2577–2590.

Grissa I, Vergnaud G, Pourcel C. 2007. The CRISPRdb database and tools to display CRISPRs and to generate dictionaries of spacers and repeats. *BMC Bioinformatics* **8:** 172.

Hsu PD, Lander ES, Zhang F. 2014. Development and applications of CRISPR–Cas9 for genome engineering. *Cell* **157:** 1262–1278.

Jinek M, Chylinski K, Fonfara I, Hauer M, Doudna JA, Charpentier E. 2012. A programmable dual-RNA-guided DNA endonuclease in adaptive bacterial immunity. *Science* **337:** 816–821.

Karvelis T, Gasiunas G, Miksys A, Barrangou R, Horvath P, Siksnys V. 2013. crRNA and tracrRNA guide Cas9-mediated DNA interference in *Streptococcus thermophilus*. *RNA Biol* **10:** 841–851.

Langmead B, Salzberg S. 2012. Fast gapped-read alignment with Bowtie 2. *Nat Methods* **9:** 357–359.

Makarova KS, Haft DH, Barrangou R, Brouns SJ, Charpentier E, Horvath P, Moineau S, Mojica FJ, Wolf YI, Yakunin AF, et al. 2011. Evolution and classification of the CRISPR–Cas systems. *Nat Rev Microbiol* **9:** 467–477.

Mali P, Yang L, Esvelt KM, Aach J, Guell M, DiCarlo JE, Norville JE, Church GM. 2013. RNA-guided human genome engineering via Cas9. *Science* **339:** 823–826.

Mojica FJ, Díez-Villaseñor C, García-Martínez J, Almendros C. 2009. Short motif sequences determine the targets of the prokaryotic CRISPR defence system. *Microbiology* **155:** 733–740.

Sander JD, Joung JK. 2014. CRISPR–Cas systems for editing, regulating and targeting genomes. *Nat Biotechnol* **32:** 347–355.

Sapranauskas R, Gasiunas G, Fremaux C, Barrangou R, Horvath P, Siksnys V. 2011. The *Streptococcus thermophilus* CRISPR/Cas system provides immunity in *Escherichia coli*. *Nucleic Acids Res* **39:** 9275–9282.

Shah SA, Erdmann S, Mojica FJ, Garrett RA. 2013. Protospacer recognition motifs: Mixed identities and functional diversity. *RNA Biol* **10:** 891–899.

Zadeh JN, Steenberg CD, Bois JS, Wolfe BR, Pierce MB, Khan AR, Dirks RM, Pierce NA. 2011. NUPACK: Analysis and design of nucleic acid systems. *J Comput Chem* **32:** 170–173.

Cite this protocol as *Cold Spring Harb Protoc*; doi:10.1101/pdb.prot086785

CHAPTER 3

Characterization of Cas9–Guide RNA Orthologs

Jonathan L. Braff,[1,4] Stephanie J. Yaung,[1,2,3,4] Kevin M. Esvelt,[1] and George M. Church[1,2,5]

[1]*Wyss Institute for Biologically Inspired Engineering, Harvard Medical School, Boston, Massachusetts 02115;* [2]*Department of Genetics, Harvard Medical School, Boston, Massachusetts 02115;* [3]*Program in Medical Engineering and Medical Physics, Harvard-MIT Division of Health Sciences and Technology, Massachusetts Institute of Technology, Cambridge, Massachusetts 02139*

In light of the multitude of new Cas9-mediated functionalities, the ability to carry out multiple Cas9-enabled processes simultaneously and in a single cell is becoming increasingly valuable. Accomplishing this aim requires a set of Cas9–guide RNA (gRNA) pairings that are functionally independent and insulated from one another. For instance, two such protein–gRNA complexes would allow for concurrent activation and editing at independent target sites in the same cell. The problem of establishing orthogonal CRISPR systems can be decomposed into three stages. First, putatively orthogonal systems must be identified with an emphasis on minimizing sequence similarity of the Cas9 protein and its associated RNAs. Second, the systems must be characterized well enough to effectively express and target the systems using gRNAs. Third, the systems should be established as orthogonal to one another by testing for activity and cross talk. Here, we describe the value of these orthogonal CRISPR systems, outline steps for selecting and characterizing potentially orthogonal Cas9–gRNA pairs, and discuss considerations for the desired specificity in Cas9-coupled functions.

INTRODUCTION

As opposed to large multisubunit complexes of Type I and III CRISPR systems, a single Cas9 protein in the Type II system can carry out the prokaryotic adaptive immune response to foreign DNA. In native prokaryotic Type II CRISPR systems, transcribed arrays are processed into CRISPR RNAs (crRNAs) that form a complex with Cas9 and a *trans*-activating RNA (tracrRNA) (Deltcheva et al. 2011). The crRNA guides Cas9 to double-stranded DNA sequences called protospacers that match the sequence of the spacer and are flanked by a protospacer-adjacent motif (PAM) unique to the CRISPR system (Mojica et al. 2009). If spacer–protospacer base-pairing is a close match, Cas9 cuts both strands of DNA.

In June of 2012, Jinek et al. first described a functional chimera of crRNA and tracrRNA called a guide RNA or gRNA (Jinek et al. 2012). The introduction of the gRNA format simplified the Type II CRISPR system from four components (target, tracrRNA, crRNA, and Cas9) to three and sparked rapid technology development in genome engineering. In December of that year, simultaneous publications from the Church and Zhang laboratories showed the chimera as a tool for multiplexable genome engineering in human cells (Cong et al. 2013; Mali et al. 2013c).

To date, numerous Cas9-mediated functions and configurations for implementation have been presented. Besides the initial applications in targeted genome editing in bacteria (Jiang et al. 2013) and eukaryotes (Cong et al. 2013; Mali et al. 2013c), Cas9 has been used for programmable activation of endogenous genes in bacteria (Bikard et al. 2012; Qi et al. 2012) and eukaryotes (Gilbert et al. 2013;

[4]These authors contributed equally to this work.

[5]Correspondence: gchurch@genetics.med.harvard.edu

Copyright © Cold Spring Harbor Laboratory Press; all rights reserved
Cite this introduction as *Cold Spring Harb Protoc*; doi:10.1101/pdb.top086793

Maeder et al. 2013; Mali et al. 2013a; Perez-Pinera et al. 2013). Cas9 can be used for epigenome regulation and visualization (Chen et al. 2013), as well as for targeted RNA degradation and localization with PAM-presenting DNA oligonucleotides (O'Connell et al. 2014). Although Cas9 naturally induces double-stranded DNA breaks, Cas9 variants can be engineered to cleave only a single strand (as a nickase) or lack endonucleolytic activity (as nuclease-null); these variants can be functionalized with other domains to recruit a variety of molecular components in a defined manner. In general, gRNAs can serve as a scaffold for assembling complexes of nucleic acids and proteins. One example is transcriptional initiation, for which there are various architectures for enhanced Cas9 activators (Chakraborty et al. 2014; Konermann et al. 2014; Chavez et al. 2015).

ORTHOGONALITY VALUE

Although powerful, a single Cas9 protein is able to mediate only one activity targeted at multiple different sites; it cannot carry out a different activity at other sites. Achieving simultaneous functionalities in a single cell requires multiple Cas9–gRNA orthologs, each engineered with a custom effector domain or modality to independently target its activity to its own array of target sites.

Given the vast potential of different Cas9-mediated functions (Mali et al. 2013b), orthogonal Cas9–gRNA pairs will be critical for independent targeting such that each function exclusively responds to its own set of gRNAs. This exclusivity insulates signal transduction pathways (Podgornaia and Laub 2013); there should be minimal, ideally zero, cross talk between Cas9–gRNA orthologs used in the same cell. Well-insulated, orthogonal biological control elements should vary independently without interfering with other components (Purnick and Weiss 2009). Orthogonal systems are further insulated by PAM specificity resulting in distinct sets of targetable protospacers. The properties of orthogonal Cas9 systems include the Cas9–gRNA interaction specificity as well as target recognition using distinct PAM sequences.

Identifying Putatively Orthogonal Cas9 Proteins

Putatively orthogonal Cas9 proteins can be selected by examining and identifying divergent repeat sequences. Tools like CRISPRfinder (Grissa et al. 2007a) and CRISPRdb (Grissa et al. 2007b) enable identification of CRISPR arrays with their constituent spacer and repeat sequences. There are also methods to experimentally validate expression and coprocessing of tracrRNAs and pre-crRNAs (Chylinski et al. 2013), and the dual RNA format can be engineered into the single gRNA format (Deltcheva et al. 2011). Candidate gRNA sequences can then be assessed to determine whether their sequences are sufficiently divergent to impart specificity to their cognate Cas9 protein. This is especially important because there is evidence for gRNA exchangeability among different Cas9s (Fonfara et al. 2013).

Characterizing Orthogonal Cas9 Proteins

PAMs are required for initial target binding, unwinding for interrogation, and subsequent cleavage of target sequences (Anders et al. 2014; Sternberg et al. 2014). PAMs are generally unique to their cognate Cas9–gRNA pair and can be identified in three ways. First, using bioinformatics tools, a multiple sequence alignment of targeted bacteriophages or plasmids can yield a consensus PAM. However, this method requires searching available sequences of phages and plasmids for matches to CRISPR spacers. It is thereby unable to predict PAMs with confidence if the availability of relevant sequences is insufficient. In addition, the bioinformatics approach may be affected by biases imparted by endogenous spacer acquisition machinery that is not generally used in engineering applications. Experimentally, in vitro cleavage assays with different plasmid substrates can produce a position weight matrix of potential PAMs, but this is relatively low-throughput compared to a library approach (Esvelt et al. 2013), which can interrogate a PAM sequence space of NNNNNNNN (corresponding to $4^8 =$ 65,536 sequences). In theory, one could perform a library selection on $4^{10} = 1{,}048{,}576$ sequences, given enough sequencing depth, if the PAM is suspected to extend out to 10 bases. For more details on

 Cite this introduction as *Cold Spring Harb Protoc*; doi:10.1101/pdb.top086793

TABLE 1. Characterized PAMs for Cas9 orthologs

Cas9 system	PAM	References	Other notes
Streptococcus thermophilus CRISPR1	NNAGAAW	Horvath et al. 2008; Esvelt et al. 2013	NNAAAAW cleaved more efficiently (Fonfara et al. 2013)
Streptococcus thermophilus CRISPR3	NGGNG	Horvath et al. 2008	
Streptococcus pyogenes	NGG	Mojica et al. 2009	
Streptococcus agalactiae	NGG	Mojica et al. 2009	
Listeria monocytogenes	NGG	Mojica et al. 2009	
Streptococcus mutans	NGG	Van der Ploeg 2009	
Neisseria meningitidis	NNNNGATT	Zhang et al. 2013; Esvelt et al. 2013	
Campylobacter jejuni	NNNNACA	Fonfara et al. 2013	
Francisella novicida	NG	Fonfara et al. 2013	
Streptococcus thermophilus LMG18311	NNGYAAA	Chen et al. 2014	NNNGYAAA seems to also work
Treponema denticola	NAAAAN	Esvelt et al. 2013	

this approach, see Protocol 1: Characterizing Cas9 Protospacer-Adjacent Motifs with High-Throughput Sequencing of Library Depletion Experiments (Braff et al. 2016).

Because the PAM sequence (Table 1) is key to distinguishing "self" versus "nonself" DNA, characterizing the PAM is important not only for Cas9-mediated activity, but also for understanding potential off-targets (Aach et al. 2014; Zhang et al. 2014). Thus, it is an essential first step to determining the orthogonality of CRISPR systems. Many groups have studied ways to improve Cas9 specificity, as Cas9–gRNA complexes are known to tolerate up to three mismatches in their targets. Notably, specificity should be evaluated with the context of the assay in mind; there are observed differences between cleavage assay results and in vivo performance in terms of PAM preference (Fonfara et al. 2013), and off-target binding does not necessarily result in off-target cleavage (Cencic et al. 2014; Wu et al. 2014). Specificity also depends on the concentration and bioavailability of the components. Factors affecting specificity include the amount of Cas9 protein, gRNA expression, and half-life of components, as well as the target and PAM sequence accessibility (Fu et al. 2013; Hsu et al. 2013; Mali et al. 2013a; Pattanayak et al. 2013). Some approaches to increase specificity include using multiple homologs for activity, finding or engineering improved variants, harnessing the dimerization requirement of FokI (Guilinger et al. 2014; Tsai et al. 2014), and using slightly truncated gRNAs (Fu et al. 2014).

CONCLUSION

Expanding the set of characterized Cas9–gRNA orthologs increases the number of engineered functions that can be simultaneously deployed in a single cell. Investigating additional Cas9 proteins can also provide insights into the nuances of varying sensitivity, specificity, and kinetic requirements in the context of the engineering application. A suite of Cas9 homologs from which to choose would allow careful matching and optimization of these parameters for a desired application. Further, the differences between PAMs found in Cas9 homologs increase the range of targetable sequences.

Undoubtedly, Type II CRISPR systems have been widely adopted for their multiplexability and programmability. The focus has been on the performance of a single function at a large number of loci and the ability to rapidly retarget this function. The increasing number of engineered variants capable of performing distinct functions has made the ability to program multiple target action pairs a very promising avenue for technology advancement.

ACKNOWLEDGMENTS

This work was supported by U.S. Department of Energy grant DE-FG02-02ER63445 (to G.M.C.) and the Wyss Institute for Biologically Inspired Engineering.

REFERENCES

Aach J, Mali P, Church G. 2014. CasFinder: Flexible algorithm for identifying specific Cas9 targets in genomes. *bioRxiv* 1–8.

Anders C, Niewoehner O, Duerst A, Jinek M. 2014. Structural basis of PAM-dependent target DNA recognition by the Cas9 endonuclease. *Nature* **513:** 569–573.

Bikard D, Hatoum-Aslan A, Mucida D, Marraffini LA. 2012. CRISPR interference can prevent natural transformation and virulence acquisition during in vivo bacterial infection. *Cell Host Microbe* **12:** 177–186.

Braff JL, Yaung SJ, Esvelt KM, Church GM. 2016. Characterizing Cas9 protospacer-adjacent motifs with high-throughput sequencing of library depletion experiments. *Cold Spring Harb Protoc* doi:10.1101/pdb.prot090183.

Cencic R, Miura H, Malina A, Robert F, Ethier S, Schmeing TM, Dostie J, Pelletier J. 2014. Protospacer adjacent motif (PAM)-distal sequences engage CRISPR Cas9 DNA target cleavage. *PLoS One* **9:** e109213.

Chakraborty S, Ji H, Kabadi A. 2014. A CRISPR/Cas9-based system for reprogramming cell lineage specification. *Stem Cell Reports* **3:** 940–947.

Chavez A, Scheiman J, Vora S, Pruitt BW, Tuttle M, P R Iyer E, Lin S, Kiani S, Guzman CD, Wiegand DJ, et al. 2015. Highly efficient Cas9-mediated transcriptional programming. *Nat Methods* **12:** 1–5.

Chen B, Gilbert LA, Cimini BA, Schnitzbauer J, Zhang W, Li G-W, Park J, Blackburn EH, Weissman JS, Qi LS, et al. 2013. Dynamic imaging of genomic loci in living human cells by an optimized CRISPR/Cas system. *Cell* **155:** 1479–1491.

Chen H, Choi J, Bailey S. 2014. Cut site selection by the two nuclease domains of the Cas9 RNA-guided endonuclease. *J Biol Chem* **289:** 13284–13294.

Chylinski K, Le Rhun A, Charpentier E. 2013. The tracrRNA and Cas9 families of type II CRISPR-Cas immunity systems. *RNA Biol* **10:** 726–737.

Cong L, Ran FA, Cox D, Lin S, Barretto R, Habib N, Hsu PD, Wu X, Jiang W, Marraffini LA, et al. 2013. Multiplex genome engineering using CRISPR/Cas systems. *Science* **339:** 819–823.

Deltcheva E, Chylinski K, Sharma CM, Gonzales K, Chao Y, Pirzada ZA, Eckert MR, Vogel J, Charpentier E. 2011. CRISPR RNA maturation by trans-encoded small RNA and host factor RNase III. *Nature* **471:** 602–627.

Esvelt KM, Mali P, Braff JL, Moosburner M, Yaung SJ, Church GM. 2013. Orthogonal Cas9 proteins for RNA-guided gene regulation and editing. *Nat Methods* **10:** 1116–1121.

Fonfara I, Le Rhun A, Chylinski K, Makarova KS, Lécrivain A-L, Bzdrenga J, Koonin EV, Charpentier E. 2013. Phylogeny of Cas9 determines functional exchangeability of dual-RNA and Cas9 among orthologous type II CRISPR-Cas systems. *Nucleic Acids Res* **42:** 2577–2590.

Fu Y, Foden JA, Khayter C, Maeder ML, Reyon D, Joung JK, Sander JD. 2013. High-frequency off-target mutagenesis induced by CRISPR-Cas nucleases in human cells. *Nat Biotechnol* **31:** 822–826.

Fu Y, Sander JD, Reyon D, Cascio VM, Joung JK. 2014. Improving CRISPR-Cas nuclease specificity using truncated guide RNAs. *Nat Biotechnol* **32:** 279–284.

Gilbert LA, Larson MH, Morsut L, Liu Z, Brar GA, Torres SE, Stern-Ginossar N, Brandman O, Whitehead EH, Doudna JA, et al. 2013. CRISPR-mediated modular RNA-guided regulation of transcription in eukaryotes. *Cell* **154:** 442–451.

Grissa I, Vergnaud G, Pourcel C. 2007a. CRISPRFinder: A web tool to identify clustered regularly interspaced short palindromic repeats. *Nucleic Acids Res* **35:** W52–W57.

Grissa I, Vergnaud G, Pourcel C. 2007b. The CRISPRdb database and tools to display CRISPRs and to generate dictionaries of spacers and repeats. *BMC Bioinformatics* **8:** 172.

Guilinger JP, Thompson DB, Liu DR. 2014. Fusion of catalytically inactive Cas9 to FokI nuclease improves the specificity of genome modification. *Nat Biotechnol.* **32:** 577–582.

Horvath P, Romero DA, Coûté-Monvoisin A-C, Richards M, Deveau H, Moineau S, Boyaval P, Fremaux C, Barrangou R. 2008. Diversity, activity, and evolution of CRISPR loci in *Streptococcus thermophilus*. *J Bacteriol* **190:** 1401–1412.

Hsu PD, Scott DA, Weinstein JA, Ran FA, Konermann S, Agarwala V, Li Y, Fine EJ, Wu X, Shalem O, et al. 2013. DNA targeting specificity of RNA-guided Cas9 nucleases. *Nat Biotechnol* **31:** 827–832.

Jiang W, Bikard D, Cox D, Zhang F, Marraffini LA. 2013. RNA-guided editing of bacterial genomes using CRISPR–Cas systems. *Nat Biotechnol* **31:** 233–239.

Jinek M, Chylinski K, Fonfara I, Hauer M, Doudna JA, Charpentier E. 2012. A programmable dual-RNA-guided DNA endonuclease in adaptive bacterial immunity. *Science* **337:** 816–821.

Konermann S, Brigham M, Trevino A. 2014. Genome-scale transcriptional activation by an engineered CRISPR–Cas9 complex. *Nature* **517:** 583–588.

Maeder ML, Linder SJ, Cascio VM, Fu Y, Ho QH, Joung JK. 2013. CRISPR RNA-guided activation of endogenous human genes. *Nat Methods* **10:** 977–979.

Mali P, Aach J, Stranges PB, Esvelt KM, Moosburner M, Kosuri S, Yang L, Church GM. 2013a. CAS9 transcriptional activators for target specificity screening and paired nickases for cooperative genome engineering. *Nat Biotechnol* **31:** 833–838.

Mali P, Esvelt KM, Church GM. 2013b. Cas9 as a versatile tool for engineering biology. *Nat Methods* **10:** 957–963.

Mali P, Yang L, Esvelt K, Aach J, Guell M. 2013c. RNA-guided human genome engineering via Cas9. *Science* **339:** 823–827.

Mojica FJM, Díez-Villaseñor C, García-Martínez J, Almendros C. 2009. Short motif sequences determine the targets of the prokaryotic CRISPR defence system. *Microbiology* **155:** 733–740.

O'Connell MR, Oakes BL, Sternberg SH, East-Seletsky A, Kaplan M, Doudna JA. 2014. Programmable RNA recognition and cleavage by CRISPR/Cas9. *Nature* **516:** 263–266.

Pattanayak V, Lin S, Guilinger JP, Ma E, Doudna JA, Liu DR. 2013. High-throughput profiling of off-target DNA cleavage reveals RNA-programmed Cas9 nuclease specificity. *Nat Biotechnol* **31:** 839–843.

Perez-Pinera P, Kocak DD, Vockley CM, Adler AF, Kabadi AM, Polstein LR, Thakore PI, Glass KA, Ousterout DG, Leong KW, et al. 2013. RNA-guided gene activation by CRISPR–Cas9-based transcription factors. *Nat Methods* **10:** 973–976.

Podgornaia AI, Laub MT. 2013. Determinants of specificity in two-component signal transduction. *Curr Opin Microbiol* **16:** 156–162.

Purnick PEM, Weiss R. 2009. The second wave of synthetic biology: From modules to systems. *Nat Rev Mol Cell Biol* **10:** 410–422.

Qi L, Haurwitz RE, Shao W, Doudna JA, Arkin AP. 2012. RNA processing enables predictable programming of gene expression. *Nat Biotechnol* **30:** 1002–1006.

Sternberg SH, Redding S, Jinek M, Greene EC, Doudna JA. 2014. DNA interrogation by the CRISPR RNA-guided endonuclease Cas9. *Nature* **507:** 62–67.

Tsai SQ, Wyvekens N, Khayter C, Foden JA, Thapar V, Reyon D, Goodwin MJ, Aryee MJ, Joung JK. 2014. Dimeric CRISPR RNA-guided FokI nucleases for highly specific genome editing. *Nat Biotechnol* **32:** 569–576.

Van der Ploeg JR. 2009. Analysis of CRISPR in *Streptococcus mutans* suggests frequent occurrence of acquired immunity against infection by M102-like bacteriophages. *Microbiology* **155:** 1966–1976.

Wu X, Scott DA, Kriz AJ, Chiu AC, Hsu PD, Dadon DB, Cheng AW, Trevino AE, Konermann S, Chen S, et al. 2014. Genome-wide binding of the CRISPR endonuclease Cas9 in mammalian cells. *Nat Biotechnol* **32:** 670–676.

Zhang Y, Heidrich N, Ampattu B. 2013. Processing-independent CRISPR RNAs limit natural transformation in *Neisseria meningitidis*. *Mol Cell* **50:** 488–503.

Zhang Y, Ge X, Yang F, Zhang L, Zheng J, Tan X, Jin Z-B, Qu J, Gu F. 2014. Comparison of non-canonical PAMs for CRISPR/Cas9-mediated DNA cleavage in human cells. *Sci Rep* **4:** 5405.

 Cite this introduction as *Cold Spring Harb Protoc*; doi:10.1101/pdb.top086793

Protocol 1

Characterizing Cas9 Protospacer-Adjacent Motifs with High-Throughput Sequencing of Library Depletion Experiments

Jonathan L. Braff,[1,4] Stephanie J. Yaung,[1,2,3,4] Kevin M. Esvelt,[1] and George M. Church[1,2,5]

[1]*Wyss Institute for Biologically Inspired Engineering, Harvard Medical School, Boston, Massachusetts 02115;*
[2]*Department of Genetics, Harvard Medical School, New Research Building, Boston, Massachusetts 02115;*
[3]*Program in Medical Engineering and Medical Physics, Harvard-MIT Division of Health Sciences and Technology, Massachusetts Institute of Technology, Cambridge, Massachusetts 02139*

This protocol outlines a general approach for characterizing the protospacer-adjacent motifs (PAMs) of Cas9 orthologs. It uses a three-plasmid system: One plasmid carries Cas9 and its tracrRNA, a second targeting vector contains the spacer and repeat, and the third plasmid encodes the targeted sequence (as the protospacer) with varying PAM sequences. It leverages the Cas9 nuclease activity to cleave and destroy plasmids that bear a compatible PAM. The level of depletion of a library of targeted plasmids after Cas9-mediated selection can then be assessed by deep sequencing to reveal candidate PAMs for downstream validation.

MATERIALS

It is essential that you consult the appropriate Material Safety Data Sheets and your institution's Environmental Health and Safety Office for proper handling of equipment and hazardous materials used in this protocol.

Reagents

Agarose gel (1%) and reagents for agarose gel electrophoresis

Antibiotics (three compatible selection agents to be used for cloning and plasmid maintenance)

Choose plasmid backbones (see below) that contain the appropriate genetic resistance markers for each selection agent.

Competent *Escherichia coli* and reagents for transformation

Use chemically competent cells for CRISPR system expression and targeting vector cloning (Step 3). For PAM library construction (Step 5) and subsequent steps, any nuclease-deficient, high-efficiency cloning strain should work. We use NEB Turbo Electrocompetent E. coli *(New England BioLabs C2986K).*

DNA for plasmid construction

CRISPR system DNA (Cas9 protein of interest, medium-strength proC constitutive promoter, and cognate tracrRNA with native promoter and terminator)

Codon-optimize the Cas9 protein sequences, if needed, using JCAT (Java codon adaptation tool; www.jcat.de) (Grote et al. 2005). The Cas9 gene and trans-activating *CRISPR RNA (tracrRNA) DNAs can be obtained via polymerase chain reaction (PCR) amplification of genomic DNA from the organism carrying the Cas9 system of interest, or synthesized by a gene synthesis service, such as gBlocks (IDT). Example DNA sequences are available in Esvelt et al. 2013. See Step 1 and Troubleshooting.*

---

[4]These authors contributed equally to this work.

[5]Correspondence: gchurch@genetics.med.harvard.edu

Copyright © Cold Spring Harbor Laboratory Press; all rights reserved

Cite this protocol as *Cold Spring Harb Protoc*; doi:10.1101/pdb.prot090183

Targeting vector DNA (strong J23100 promoter, unique 20-nt spacer sequence, and repeat sequence for the CRISPR system of interest)

Select spacer sequences that will be functional in Escherichia coli *(for two example sequences, see Esvelt et al. 2013). At least two versions of the targeting construct are required to implement proper controls; see Step 2.*

Glycerol (10% v/v in ddH_2O), sterile

Isothermal assembly mix

Use Gibson Assembly Master Mix (New England BioLabs E2611S) or prepare isothermal assembly mix in the laboratory as described by Gibson et al. (2009).

LB liquid medium

In addition, prepare plates containing LB medium solidified with agar.

PCR mix

Use a high-fidelity polymerase with proofreading activity such as KAPA HiFi HotStart ReadyMix PCR Kit (Kapa Biosystems) or Phusion polymerase (New England BioLabs). Use nuclease-free water for PCR.

Plasmid backbones for construction of three-plasmid system

Each vector type (the CRISPR system expression vector, the targeting vectors, and the PAM libraries) will need to coexist in a single cell with two other types of plasmids, so select plasmid backbones with appropriately compatible origins and selection markers. For example, the three plasmids can be p15a origin with chloramphenicol resistance, colE1 orgin with kanamycin resistance, and cloD13 origin with spectinomycin resistance. Plasmids can be obtained from Addgene.

Reagents for construction of protospacer-adjacent motif (PAM) libraries (see Step 4)

Sequencing reagents (Illumina MiSeq)

Equipment

Equipment for agarose gel electrophoresis

Incubator at 37°C

Microcentrifuge

PCR tubes

Sequencing platform (Illumina MiSeq)

Shaking incubator at 37°C

Spin columns for PCR purification, plasmid DNA preparation, and plasmid mini-prep (QIAGEN)

Thermal cycler

METHOD

Constructing Plasmids

1. Using isothermal assembly (Gibson et al. 2009), create the CRISPR system expression vector containing the Cas9 protein of interest preceded by a medium-strength proC constitutive promoter, immediately followed by its cognate tracrRNA with its native promoter and terminator.

   See Troubleshooting.

2. Using isothermal assembly, create at least two versions of the targeting vector, each containing a strong J23100 promoter followed by a unique 20-nt spacer sequence immediately adjacent to the repeat sequence for the CRISPR system of interest.

   The two versions of the targeting vector will not need to coexist in a cell together, so they can and should contain identical plasmid backbones. This design controls for any biases in PAM library construction and selection by allowing the comparison between library depletion in a properly targeted (i.e., spacer matches the protospacer) experiment versus library depletion in a nontargeted (i.e., spacer does not match the protospacer) experiment.

Cite this protocol as *Cold Spring Harb Protoc*; doi:10.1101/pdb.prot090183

3. Transform each of the three assemblies into a chemically competent *E. coli* cloning strain. Grow cells overnight on LB agar plates with the appropriate antibiotic for selection. Verify the sequences of the selected clones and perform a plasmid DNA preparation of correct clones.
4. Construct two PAM libraries (one for each of the targeting plasmids), each containing a protospacer that matches one of the spacers from Step 2, followed by an 8-bp degenerate region immediately 3′ of the spacer where the PAM site would typically be located.

   Standard PCR and isothermal assembly methods can be used; for example, we amplify a backbone vector such as pZE21 (ExpressSys) using primers encoding one of the two protospacer sequences followed by eight random bases. We typically split the vector into two pieces, and re-assemble the pieces (now with the added protospacer and PAM sequences) in isothermal assembly.
5. Transform the libraries into a high-efficiency cloning strain. Grow the cells overnight in at least 500 mL of LB liquid medium with the appropriate antibiotic for selection, and perform a plasmid DNA preparation from each library.

   When cloning libraries, it is important to plate a few different dilutions of the cells to accurately quantitate the number of transformants and ensure adequate coverage of the library. Aim to have ~1×10^8 transformants. Simple cloning optimizations and/or more than one transformation of each library may be needed to achieve this library size.

   It may be desirable to prepare the plasmid at a slightly larger scale, as libraries can be reused to characterize multiple orthologs.
6. Prepare a total of four recipient strains (each containing a targeting plasmid in the presence of the CRISPR expression plasmid) for pairing with the PAM library (Step 7) as follows.
   i. Cotransform the CRISPR system expression plasmid and one targeting plasmid into a nuclease-deficient, high-efficiency cloning strain.
   ii. Repeat Step 6.i using the second targeting plasmid.
   iii. Grow the cells from each transformation overnight on LB agar plates containing both antibiotics for selection (i.e., one for each plasmid resistance marker).
   iv. Perform two cell preparations from each strain by pelleting 50 mL of a monoclonal liquid culture in late exponential growth, washing the cells three times with 50 mL of 10% glycerol, and then resuspending the cells in 50 µL of ddH_2O (double-distilled water).
7. Transform each sample from Step 6 with 200 ng of PAM library DNA, such that each of the targeting strains is paired with each library, for a total of four transformations. Recover the cells for 2 h at 37°C.
8. Dilute the cells 100× in LB medium containing all three appropriate antibiotics for selection (i.e., one for each plasmid resistance marker) and grow the cells overnight (12–18 h). In addition, plate several dilutions of the recovery culture from Step 7 to estimate posttransformation library size.

   More than 1×10^7 clones corresponds to complete coverage of 65,536 (4^8) PAM sequences.

Performing High-Throughput Sequencing and Analysis

9. Select library clones after 12–18 h of antibiotic selection. Harvest DNA from each of the selections by plasmid mini-prep.
10. Prepare an Illumina sequencing library via PCR amplification of the degenerate region of the PAM library, and sequence using MiSeq.

    It is important to consult the sequencing core or Illumina directly regarding current best practice for amplicon sequencing to ensure the amplicon library has sufficient sequence diversity, appropriate adapters, and proper indexing to demultiplex the various selections after sequencing. Paired-end 30-bp reads with custom sequencing primers work well. If custom primers will be used, make sure the melting temperature is high enough for the sequencing platform.
11. Preprocess the sequencing results by merging the paired-end reads and filtering out any pairs without perfect alignment, the protospacer sequences, and the plasmid backbones. Trim the filtered reads, leaving only the degenerate PAM regions.

12. Calculate fold depletion for each candidate PAM by comparing the position weight matrix (PWM) of the spacer–protospacer matching pool with that of the mismatching pool while filtering for the candidate PAM sequence.

    A simple Python script called patternProp3.py is helpful for this step (see Esvelt et al. 2013; supplemental). A list of candidate PAMs can be generated exhaustively or by inspection. To help generate the list by inspection, patternProp3.py outputs not only the total fold depletion for a candidate PAM but also the specific depletion for each base at each position for all reads matching that candidate. One can first look at the fold depletions of the null filter (i.e., NNNNNNNN), and then identify the most depleted bases and use those as the new filter. It is important to be aware that as the filter becomes more complex, there are fewer reads, so the differences may become statistically insignificant.

    See Troubleshooting.

(Optional) Validating PAM Sequences

13. To validate a putative PAM identified in the experiment, redesign the library plasmid to encode the PAM and measure depletion.

    Alternatively, validation can be completed with a GFP (green fluorescent protein) reporter plasmid, in which the protospacer and PAM are in the 5′ UTR (untranslated region) of GFP (which has a strong pR promoter and T7 g10 ribosome binding site), as described by Esvelt et al. (2013).

TROUBLESHOOTING

Problem (Step 1): The experiment requires bacterial expression of novel orthologs, yet performing a PAM-finding experiment requires a priori knowledge of the Cas9 ortholog of interest, specifically the corresponding sequence and directionality of the tracrRNA and crRNA.

Solution: Use tools such as CRISPRfinder (Grissa et al. 2007). Include the native promoter and terminator of the tracrRNA, and place the cassette downstream from Cas9. Minimize the secondary structure of the spacer and repeat.

Problem (Step 12): Depletion biases are present in the library experiment.

Solution: There can be apparent minimum and maximum fold depletion values (Esvelt et al. 2013). Some are low-frequency mutations leading to inactivated Cas9 or target escape. Analysis should focus on moderately depleted sequences. Changing the recovery time may vary the stringency of the selection. In addition, the library selection can be performed in the context of the final activity of interest (i.e., the protocol described here would be for cleavage activity).

ACKNOWLEDGMENTS

This work was supported by U.S. Department of Energy grant DE-FG02-02ER63445 (to G.M.C.) and the Wyss Institute for Biologically Inspired Engineering.

REFERENCES

Esvelt KM, Mali P, Braff JL, Moosburner M, Yaung SJ, Church GM. 2013. Orthogonal Cas9 proteins for RNA-guided gene regulation and editing. *Nat Methods* **10:** 1116–1121.

Gibson DG, Young L, Chuang R-Y, Venter JC, Hutchison CA, Smith HO. 2009. Enzymatic assembly of DNA molecules up to several hundred kilobases. *Nat Methods* **6:** 343–345.

Grissa I, Vergnaud G, Pourcel C. 2007. CRISPRFinder: A web tool to identify clustered regularly interspaced short palindromic repeats. *Nucleic Acids Res* **35:** W52–W57.

Grote A, Hiller K, Scheer M, Münch R, Nörtemann B, Hempel DC, Jahn D. 2005. JCat: A novel tool to adapt codon usage of a target gene to its potential expression host. *Nucleic Acids Res* **33:** W526–W531.

 Cite this protocol as *Cold Spring Harb Protoc*; doi:10.1101/pdb.prot090183

CHAPTER 4

Large-Scale Single Guide RNA Library Construction and Use for CRISPR–Cas9-Based Genetic Screens

Tim Wang,[1,2,3,4,5] Eric S. Lander,[1,3,6,7,8] and David M. Sabatini[1,2,3,4,5,7,8]

[1]*Department of Biology, Massachusetts Institute of Technology, Cambridge, Massachusetts 02139;* [2]*Whitehead Institute for Biomedical Research, Cambridge, Massachusetts 02142;* [3]*Broad Institute of MIT and Harvard, Cambridge, Massachusetts 02142;* [4]*David H. Koch Institute for Integrative Cancer Research at MIT, Cambridge, Massachusetts 02139;* [5]*Howard Hughes Medical Institute, Department of Biology, Massachusetts Institute of Technology, Cambridge, Massachusetts 02139;* [6]*Department of Systems Biology, Harvard Medical School, Boston, Massachusetts 02115*

The ability to systematically disrupt genes serves as a powerful tool for understanding their function. The programmable CRISPR–Cas9 system enables efficient targeting of large numbers of genes through the use of single guide RNA (sgRNA) libraries. In cultured mammalian cells, collections of knockout mutants can be readily generated by means of transduction of Cas9–sgRNA lentiviral pools, screened for a phenotype of interest, and counted using high-throughput DNA sequencing. This technique represents the first general method for undertaking systematic loss-of-function genetic screens in mammalian cells. Here, we introduce the methodology and rationale for conducting CRISPR-based screens, focusing on distinguishing positive and negative selection strategies.

INTRODUCTION

The molecular underpinnings of many fundamental cellular pathways have been deciphered through unbiased genetic screens in microorganisms. However, similar studies in human cells have been hampered by a lack of suitable tools for manipulating their large, diploid genomes, and this has limited our understanding of the genes and biological processes unique to mammals. Recently, the bacterial CRISPR–Cas9 adaptive immune system has been co-opted to enable efficient, sequence-specific DNA cleavage in cultured cells and whole organisms, greatly expanding the toolbox for mammalian geneticists (Cong et al. 2013; Mali et al. 2013; Wang et al. 2013). In contrast to previous genome-editing techniques, targeting reagents for the CRISPR–Cas9 system can be rapidly generated as the target specificity is dictated by a short 20-bp sequence at the 5′ end of the sgRNA. This ease of construction allows the generation of large-scale libraries targeting all (or a desired subset) of the protein-coding genes encoded in a mammalian genome by using microarray-based oligonucleotide synthesis. Using this approach, we and others have developed a general method for performing systematic loss-of-function genetic screens in mammalian cells (Shalem et al. 2014; Wang et al. 2014).

Below, we summarize the steps required to carry out a CRISPR-based screen. Additional considerations relating to the design of screens and validation of hits will not be discussed at length here. For these topics, we refer the reader to articles published elsewhere (Moffat and Sabatini 2006; Boutros and

---

[7]These authors contributed equally to this work.

[8]Correspondence: lander@broadinstitute.org; sabatini@wi.mit.edu

Copyright © Cold Spring Harbor Laboratory Press; all rights reserved

Cite this introduction as *Cold Spring Harb Protoc*; doi:10.1101/pdb.top086892

Ahringer 2008; Kaelin 2012). In associated protocols, we present details for designing and preparing sgRNA libraries suitable for genetic screening (see Protocol 1: Single Guide RNA Library Design and Construction [Wang et al. 2016a]) and then give a detailed method for their use in lentiviral packaging, followed by infecting and screening of a cell line of interest, together with recommendations for data-analysis options (see Protocol 2: Viral Packaging and Cell Culture for CRISPR-Based Screens [Wang et al. 2016b]).

SCREEN PRINCIPLE

The bacterial CRISPR–Cas9 system has been co-opted for mammalian genome editing, allowing for the rapid generation of isogenic cell lines and mice with modified alleles. By using pooled libraries expressing tens of thousands of sgRNAs, the scale of this technology can be greatly expanded, enabling loss-of-function genetic screening of all protein-coding genes in mammalian cells. In this method, sgRNA expression constructs are generated by array-based oligonucleotide library synthesis and packaged into lentiviral particles (Fig. 1). Target cells of interest can then be transduced with the lentiviral sgRNA pools to generate a collection of knockout mutants through Cas9-mediated genomic cleavage. Finally, through high-throughput sequencing of the integrated expression cassettes, the number of cells bearing each sgRNA in the mutant collection can be monitored over time to pinpoint the mutants of interest.

POSITIVE SELECTION SCREENS

Pooled screens can be divided into two classes, positive selection and negative selection, which can often reveal complementary biological information (Fig. 2). In positive selection screens, disruption of the genes of interest confers a selective advantage on cells, allowing them to rise to high frequency. As a result, gene candidates can be readily identified.

Some biological processes, such as drug resistance or anchorage-independent growth, are ideally suited for positive selection screening as they are intrinsically linked to cellular proliferation and survival. For studying other processes, additional selection strategies can be devised. For example, cells can be engineered to express a selectable marker in a pathway-activity-dependent manner or

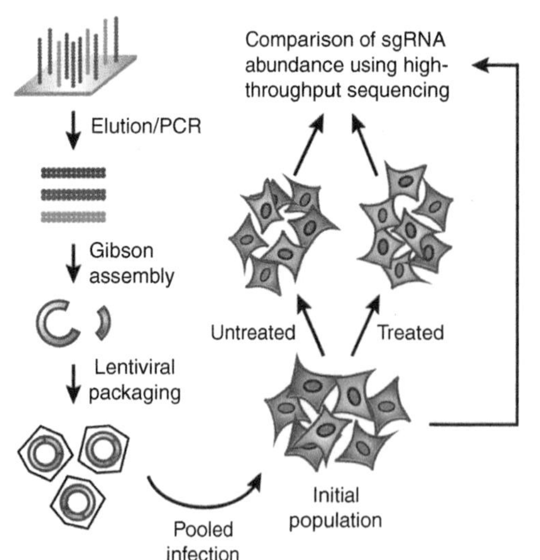

FIGURE 1. Schematic of sgRNA library construction and genetic screening strategy. Illustrated is a summary of the steps required to generate sgRNA expression constructs by array-based oligonucleotide library synthesis that are subsequently packaged into lentiviral particles. The initial population of target cells is transduced with the lentiviral sgRNA pool to generate a collection of knockout mutants by Cas9-mediated genomic cleavage. Finally, high-throughput sequencing enables comparison of the integrated expression cassettes, and the number of cells bearing each sgRNA in the mutant collection can be monitored over time in treated versus untreated cells.

Cite this introduction as *Cold Spring Harb Protoc*; doi:10.1101/pdb.top086892

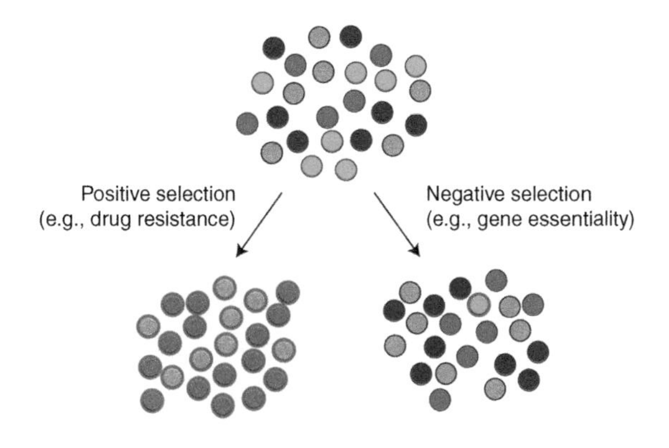

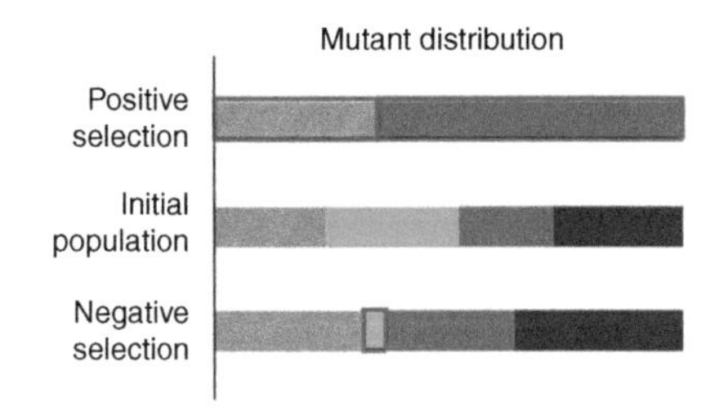

FIGURE 2. Positive and negative selection screens. The mutants of interest (outlined in red) rise to high frequency in positive selection screens but are underrepresented in negative selection screens, and consequently require more-precise methods for their detection.

isolated in screens using fluorescence-activated cell sorting (Duncan et al. 2012; Lee et al. 2013). Together, these approaches can greatly broaden the diversity of phenotypes amenable for screening.

NEGATIVE SELECTION SCREENS

Negative selection screens seek to identify genes whose inactivation is detrimental to cells. Such genes can be recognized by a decrease in the abundance of corresponding sgRNAs during the course of a screen. Although conceptually simple, identifying such sgRNAs poses a significant technical challenge for pooled CRISPR-based screens. First, negative selection screens require potent sgRNAs, because depletion of an sgRNA can only be observed if the gene target is cleaved and inactivated in a large proportion of the cells carrying the sgRNA. Additionally, during the infection of the library, it is necessary to introduce each sgRNA into a large number (~1000) of target cells. This high level of representation of the library in the initial cell population ensures that a "drop-out" of cells carrying a deleterious sgRNA can be reliably distinguished from random changes in abundance resulting from sampling fluctuations. For genome-wide screens, particularly in primary cells or in vivo, it can be impractical or impossible to infect and culture cells at the required scale. In these cases, secondary screens using a sublibrary of sgRNAs targeting candidate genes can serve as a powerful tool for validation.

ACKNOWLEDGMENTS

Our work was supported by the U.S. National Institutes of Health (CA103866) (D.M.S.), National Human Genome Research Institute (2U54HG003067-10) (E.S.L.), the Broad Institute of the Massachusetts Institute of Technology and Harvard (E.S.L.), and an award from the U.S. National Science Foundation (T.W.). T.W, E.S.L., and D.M.S. are inventors on a patent application relating to aspects of work described here.

REFERENCES

Boutros M, Ahringer J. 2008. The art and design of genetic screens: RNA interference. *Nat Rev Genet* **9:** 554–566.

Cong L, Ran FA, Cox D, Lin S, Barretto R, Habib N, Hsu PD, Wu X, Jiang W, Marraffini LA, et al. 2013. Multiplex genome engineering using CRISPR/Cas systems. *Science* **339:** 819–823.

Duncan LM, Timms RT, Zavodszky E, Cano F, Dougan G, Randow F, Lehner PJ. 2012. Fluorescence-based phenotypic selection allows forward genetic screens in haploid human cells. *PLoS One* **7:** e39651.

Kaelin WG. 2012. Use and abuse of RNAi to study mammalian gene function. *Science* **337:** 421–422.

Lee CC, Carette JE, Brummelkamp TR, Ploegh HL. 2013. A reporter screen in a human haploid cell line identifies CYLD as a constitutive inhibitor of NF-κB. *PLoS One* **8:** e70339.

Mali P, Yang L, Esvelt KM, Aach J, Guell M, DiCarlo JE, Norville JE, Church GM. 2013. RNA-guided human genome engineering via Cas9. *Science* **339:** 823–826.

Moffat J, Sabatini DM. 2006. Building mammalian signalling pathways with RNAi screens. *Nat Rev Mol Cell Biol* **7:** 177–187.

Shalem O, Sanjana NE, Hartenian E, Shi X, Scott DA, Mikkelsen TS, Heckl D, Ebert BL, Root DE, Doench JG, et al. 2014. Genome-scale CRISPR-Cas9 knockout screening in human cells. *Science* **343:** 84–87.

Wang H, Yang H, Shivalila CS, Dawlaty MM, Cheng AW, Zhang F, Jaenisch R. 2013. One-step generation of mice carrying mutations in multiple genes by CRISPR/Cas-mediated genome engineering. *Cell* **153:** 910–918.

Wang T, Wei JJ, Sabatini DM, Lander ES. 2014. Genetic screens in human cells using the CRISPR-Cas9 system. *Science* **343:** 80–84.

Wang T, Lander ES, Sabatini DM. 2016a. Single guide RNA library design and construction. *Cold Spring Harb Protoc* doi: 10.1101/pdb.prot090803.

Wang T, Lander ES, Sabatini DM. 2016b. Viral packaging and cell culture for CRISPR-based screens. *Cold Spring Harb Protoc* doi: 10.1101/pdb.prot090811.

Cite this introduction as *Cold Spring Harb Protoc*; doi:10.1101/pdb.top086892

Protocol 1

Single Guide RNA Library Design and Construction

Tim Wang,[1,2,3,4,5] Eric S. Lander,[1,3,6,7,8] and David M. Sabatini[1,2,3,4,5,7,8]

[1]*Department of Biology, Massachusetts Institute of Technology, Cambridge, Massachusetts 02139;* [2]*Whitehead Institute for Biomedical Research, Cambridge, Massachusetts 02142;* [3]*Broad Institute of MIT and Harvard, Cambridge, Massachusetts 02142;* [4]*David H. Koch Institute for Integrative Cancer Research at MIT, Cambridge, Massachusetts 02139;* [5]*Howard Hughes Medical Institute, Department of Biology, Massachusetts Institute of Technology, Cambridge, Massachusetts 02139;* [6]*Department of Systems Biology, Harvard Medical School, Boston, Massachusetts 02115*

This protocol describes how to generate a single guide RNA (sgRNA) library for use in genetic screens. There are many online tools available for predicting sgRNA sequences with high target specificity and/or cleavage activity. Here, we refer the user to genome-wide sgRNA sequence predictions that we have developed for both the human and mouse and that are available from the Broad Institute website. Once a set of target genes and corresponding sgRNA sequences has been identified, customized oligonucleotide pools can be rapidly synthesized by a number of commercial vendors. Thereafter, as described here, the oligonucleotides can be efficiently cloned into an appropriate lentiviral expression vector backbone. The resulting plasmid pool can then be packaged into lentiviral particles and used to generate knockouts in any cell line of choice.

MATERIALS

It is essential that you consult the appropriate Material Safety Data Sheets and your institution's Environmental Health and Safety Office for proper handling of equipment and hazardous materials used in this protocol.

RECIPES: Please see the end of this protocol for recipes indicated by <R>. Additional recipes can be found online at http://cshprotocols.cshlp.org/site/recipes.

Reagents

Agarose gels (1.0% and 2.0%)
BsmBI (New England Biolabs R0580S)
Endura Electrocompetent Cells (Lucigen 60242-0)
Ethidium bromide
Gel Extraction Kit (QIAGEN 28704)
Gibson Assembly Master Mix (New England Biolabs E2611S)
LB-ampicillin agar plates <R>
LB (Luria–Bertani) liquid medium <R>
Lentiviral single guide RNA (sgRNA) expression plasmid

Two expression plasmids are suitable—lentiCRISPR v2 (sgRNA expression plasmid with Cas9 [Addgene 52961]) or lentiGuide-Puro (sgRNA expression plasmid without Cas9 [Addgene 52963]). See Discussion for further details.

---

[7]These authors contributed equally to this work.
[8]Correspondence: lander@broadinstitute.org; sabatini@wi.mit.edu

Copyright © Cold Spring Harbor Laboratory Press; all rights reserved
Cite this protocol as *Cold Spring Harb Protoc*; doi:10.1101/pdb.prot090803

Library polymerase chain reaction (PCR) primers
Forward: GGCTTTATATATCTTGTGGAAAGGACGAAACACCG
Reverse: CTAGCCTTATTTTAACTTGCTATTTCTAGCTCTAAAAC

NEBuffer 3.1 (New England Biolabs B7203S, supplied by manufacturer as a 10× stock together with BsmBI)
Oligonucleotides, custom-made/ordered (see Step 4)
Phusion High-Fidelity PCR Master Mix with HF Buffer (New England Biolabs M0531S)
Plasmid Plus Maxi Kit (QIAGEN 12963)
Recovery medium (Lucigen 80026-1)
Water (H_2O), PCR-grade

Equipment

Bacterial shaker(s) (at 30°C and 37°C)
E. coli Pulser Transformation Apparatus (Bio-Rad 165-2101)
Eppendorf tubes (1.5-mL)
Erlenmeyer flask (500-mL)
Gel electrophoresis apparatus
Gel imager
Heat blocks (at 50°C and 55°C)
Ice
Incubator(s) (at 30°C and 37°C)
MicroPulser Cuvettes (Bio-Rad 165-2089)
NanoDrop spectrophotometer (NanoDrop)
Online sgRNA sequence analysis tools (see Step 1)
Pipette
Thermocycler
Water bath
x-tracta gel extractor (USA Scientific 5454-0100)

METHOD

This protocol is intended primarily for users who wish to construct an sgRNA library targeting a customized set of genes. Many large-scale sgRNA libraries suitable for Cas9-based screening can be found on Addgene (http://www.addgene.org/CRISPR/libraries/). If a preexisting library is used, the investigator should skip to the Library Transformation section of this protocol (Steps 24–30).

sgRNA Sequence Design

1. Obtain a list of sgRNA sequences targeting the genes of interest.

   Investigators have a wide choice of online tools for determining sgRNA sequences that possess high target specificity and/or cleavage activity (Heigwer et al. 2014; Xie et al. 2014). For human and mouse genes, we have generated a set of sgRNA sequences that can be accessed at http://www.broadinstitute.org/~timw/CRISPR/. These sets of sgRNA predictions have been experimentally validated to show high on-target cleavage activity (T. Wang, unpubl.), and we recommend their use here.

2. Prepend the 5′ universal flanking sequence: TATCTTGTGGAAAGGACGAAACACC.

   An additional "G" must be prepended to sgRNA sequences starting with any other nucleotide to allow efficient transcription from the U6 promoter.

3. Append the 3′ universal flanking sequence: GTTTTAGAGCTAGAAATAGCAAGTTAAAAT.

4. Order custom oligonucleotide pools.

   Microarray-based oligonucleotide synthesis is a highly competitive and rapidly evolving industry, and, as such, many commercial vendors can provide similar product offerings. Many of the sgRNA libraries created to date have been synthesized by CustomArray Inc. (Bothell, WA), although we do recommend that the user identify a suitable vendor depending on the desired scale, accuracy, and speed of synthesis.

Cite this protocol as *Cold Spring Harb Protoc*; doi:10.1101/pdb.prot090803

Vector Preparation

5. Streak out a bacterial stab culture of sgRNA lentiviral expression vector obtained from Addgene on LB-amp plates and incubate overnight at 30°C.
6. Pick a single colony and seed into a 500-mL Erlenmeyer flask containing 100 mL of LB liquid medium containing 100 µg/mL ampicillin.
7. Incubate culture overnight at 30°C in a rotating bacterial shaker.
8. Prepare plasmid DNA from the bacterial culture using the QIAGEN Plasmid Plus Maxi Kit according to the manufacturer's instructions.
9. Assemble the following digestion reaction on ice.

Lentiviral sgRNA expression plasmid	3 µg
NEBuffer 3.1	3 µL
BsmBI	3 µL
H_2O	to 30 µL

10. Incubate overnight at 55°C in a water bath.
11. Run out the reaction on an ethidium-bromide-stained 1% agarose gel. Visualize the digested bands using a standard gel imager.
12. Cut the digested vector backbone using an x-tracta gel extractor tool.
13. Extract DNA using the QIAGEN Gel Extraction Kit according to the manufacturer's instructions, eluting in 10 µL of water.

Library Amplification and Cloning

14. Assemble four replicates of the following PCR on ice, as follows.

Synthesized oligonucleotides	1 µL
Forward library PCR primer (10 µM)	2 µL
Reverse library PCR primer (10 µM)	2 µL
Phusion High-Fidelity PCR Master Mix with HF Buffer	25 µL
H_2O	20 µL

15. Amplify reactions in a thermocycler using the following program, varying the total number of cycles for each replicate.

1 cycle	98°C	2 min
8, 10, 12 or 16 cycles	98°C	10 sec
	60°C	15 sec
	72°C	45 sec
1 cycle	72°C	5 min
1 cycle	4°C	Hold

16. Run out the reactions on an ethidium-bromide-stained 2% agarose gel. Visualize the PCR bands using a standard gel imager.
17. For all reactions yielding a visible product at 92 base pairs, cut out the band using an x-tracta gel extractor tool.
18. Extract DNA using the QIAGEN Gel Extraction Kit according to the manufacturer's instructions, eluting in 10 µL of water.
19. Determine the PCR product concentrations using a NanoDrop spectrophotometer. Proceed to Gibson Assembly cloning using the sample amplified for the fewest cycles, with a product concentration >10 ng/µL.

20. Assemble two replicates of the following Gibson Assembly reaction on ice.

Digested vector from Step 13	100 ng
Gibson Assembly Master Mix	10 µL
H_2O	to 19 µL

21. Add 1 µL of the library PCR product to one reaction and add 1 µL of water to the other.
22. Incubate for 1 h at 50°C.
23. Place reactions on ice after completion.

Library Transformation

24. Warm Recovery Medium for 30 min in a 37°C water bath.
25. Warm an LB-ampicillin agar plate for 30 min in a 37°C incubator.
26. Thaw one vial of Endura Electrocompetent Cells and aliquot cells into two tubes on ice for 15 min.
27. Place two MicroPulser Cuvettes on ice.
28. For each reaction (control- and insert-containing) proceed as follows:

    i. Add 1 µL of the Gibson Assembly reaction product to bacterial cells.

    ii. Transfer 25 µL of the bacterial cell and Gibson Assembly reaction product mixture into MicroPulser Cuvettes.

    iii. Place cuvette into an *Escherichia coli* Pulser Transformation Apparatus and electroporate at 1.8 kV.

    iv. Quickly add 975 µL of the Recovery Medium into the cuvette and pipette up and down three times to resuspend the cells.

    v. Transfer mixture to a 1.5-mL microcentrifuge tube.

    vi. Place the tube in a shaking incubator for 1 h at 37°C.

    vii. Serially dilute 10 µL of the transformation mixture in Recovery Medium four times, using a dilution factor of 1/10 at each step.

    viii. Spot 10 µL of each dilution onto an LB-ampicillin plate.

    ix. Incubate plate overnight at 30°C.

    The number of colonies on these spots can be multiplied by 10^3, 10^4, 10^5, and 10^6, respectively, to estimate the total number of colony-forming units.

    See Troubleshooting.

29. For insert-containing reaction only, proceed as follows:

    i. Seed the remainder of the transformation mixture into a 500-mL Erlenmeyer flask containing 100 mL of LB liquid medium containing 100 µg/mL ampicillin.

    ii. Incubate culture overnight at 30°C.

    iii. If the transformation efficiency, as assessed by the serial plating, exceeds 20-fold of the library size and the transformation efficiency of the control reaction is <1% of the insert-containing reaction, then prepare plasmid DNA from the bacterial culture using the QIAGEN Plasmid Plus Maxi Kit according to the manufacturer's instructions.

30. To assess recombination, run out the amplified plasmid on an ethidium-bromide-stained 1% agarose gel. Visualize the plasmid DNA using a standard gel imager.

    See Troubleshooting.

Cite this protocol as *Cold Spring Harb Protoc*; doi:10.1101/pdb.prot090803

TROUBLESHOOTING

Problem (Step 28): The transformation efficiency is too low.
Solution: There are two common causes of this problem.

- There are bad electrocompetent cells. To check, perform a test electroporation with an intact control plasmid and compare with the advertised efficiency.
- The level of salt in the transformation is too high. Dilute the Gibson Assembly reactions 1:3 in water before transforming.

Finally, monitoring the time constant after electroporating cells can often serve as a useful indicator of transformation efficiency. A time constant between 3.5 and 4.5 msec is ideal.

Problem (Step 30): The plasmid library is recombined.
Solution: The use of Endura cells and incubation of bacteria at 30°C are both intended to minimize recombination of the lentiviral plasmid library. However, if a substantial fraction of the amplified plasmid library is recombined, as assessed by gel electrophoresis, it might be advisable to grow the transformation products on agar plates rather than in liquid culture.

DISCUSSION

The decision to use a vector with or without Cas9 for screening depends on several factors. Using a Cas9-containing backbone readily allows screening in any cell line without prior modification of the cell line. However, much less recombination during plasmid amplification and higher viral titers (typically 20- to 100-fold higher) during viral packaging can be achieved by using smaller vectors lacking Cas9. For this reason, we recommend that only users who plan to conduct screens across multiple cell lines should clone sgRNA libraries into a Cas9-containing vector. In contrast, those seeking to perform screens across multiple conditions in a single cell line should first derive a Cas9-expressing clone.

RECIPES

Ampicillin Stock Solution (100 mg/mL)

Ampicillin (sodium salt [sodium ampicillin], m.w. = 371.40)

Dissolve 1 g of sodium ampicillin in sufficient H_2O to make a final volume of 10 mL. If sterilization is required, prewash a 0.45- or 0.22-µm sterile filter by drawing through 50–100 mL of H_2O. Then pass the ampicillin solution through the washed filter. Store the ampicillin in aliquots at −20°C for 1 yr (or at 4°C for 3 mo).

LB Agar

Agar (20 g/L)
NaCl (10 g/L; Sigma-Aldrich S9625)
Tryptone (10 g/L; BD 211705)
Yeast extract (5 g/L; BD 212750)

Add H_2O to a final volume of 1 L. Adjust the pH to 7.0 with 5 N NaOH. Autoclave. Pour into Petri dishes (~ 25 mL per 100-mm plate).

LB-Ampicillin Agar Plates

Ampicillin, filter-sterilized (10 mg/mL stock)
LB agar

Autoclave 1 L of LB agar. Cool to 55°C. Add 10 mL of ampicillin stock. Pour into Petri dishes (~ 25 mL per 100-mm plate).

LB (Luria-Bertani) Liquid Medium

Reagent	Amount to add
H_2O	950 mL
Tryptone	10 g
NaCl	10 g
Yeast extract	5 g

Combine the reagents and shake until the solutes have dissolved. Adjust the pH to 7.0 with 5 N NaOH (~0.2 mL). Adjust the final volume of the solution to 1 L with H_2O. Sterilize by autoclaving for 20 min at 15 psi (1.05 kg/cm^2) on liquid cycle.

REFERENCES

Heigwer F, Kerr G, Boutros M. 2014. E-CRISP: Fast CRISPR target site identification. *Nat Methods* **11:** 122–123.

Xie S, Shen B, Zhang C, Huang X, Zhang Y. 2014. sgRNAcas9: A software package for designing CRISPR sgRNA and evaluating potential off-target cleavage sites. *PLoS One* **9:** e100448.

Cite this protocol as *Cold Spring Harb Protoc*; doi:10.1101/pdb.prot090803

Protocol 2

Viral Packaging and Cell Culture for CRISPR-Based Screens

Tim Wang,[1,2,3,4,5] Eric S. Lander,[1,3,6,7,8] and David M. Sabatini[1,2,3,4,5,7,8]

[1]*Department of Biology, Massachusetts Institute of Technology, Cambridge, Massachusetts 02139;* [2]*Whitehead Institute for Biomedical Research, Cambridge, Massachusetts 02142;* [3]*Broad Institute of MIT and Harvard, Cambridge, Massachusetts 02142;* [4]*David H. Koch Institute for Integrative Cancer Research at MIT, Cambridge, Massachusetts 02139;* [5]*Howard Hughes Medical Institute, Department of Biology, Massachusetts Institute of Technology, Cambridge, Massachusetts 02139;* [6]*Department of Systems Biology, Harvard Medical School, Boston, Massachusetts 02115*

This protocol describes how to perform the tissue culture and high-throughput sequencing library preparation for a CRISPR-based screen. First, pantropic lentivirus is prepared from a single guide RNA (sgRNA) plasmid pool and applied to the target cells. Following antibiotic selection and a harvest of the initial population, cells are then cultured under the desired screening condition(s) for 14 population doublings. The sgRNA barcode sequences integrated in the genomic DNA of each cell population are amplified and subject to high-throughput sequencing. Guidelines for downstream analysis of the sequencing data are also provided.

MATERIALS

It is essential that you consult the appropriate Material Safety Data Sheets and your institution's Environmental Health and Safety Office for proper handling of equipment and hazardous materials used in this protocol.

RECIPES: Please see the end of this protocol for recipes indicated by <R>. Additional recipes can be found online at http://cshprotocols.cshlp.org/site/recipes.

Reagents

Agarose gel (1.0%)
Cell-culture medium appropriate for target cells
Dulbecco's Modified Eagle Medium (DMEM), high-glucose, GlutaMAX Supplement (Gibco 10566-016)
Ethidium bromide
Fetal bovine serum (inactivated) (Sigma-Aldrich F4135)
Gel Extraction Kit (QIAGEN 28704)
Human embryonic kidney (HEK) 293 T cells (ATCC CRL-3216)
LB-ampicillin agar plates <R>

[7]These authors contributed equally to this work
[8]Correspondence: lander@broadinstitute.org; sabatini@wi.mit.edu

Copyright © Cold Spring Harbor Laboratory Press; all rights reserved
Cite this protocol as *Cold Spring Harb Protoc*; doi:10.1101/pdb.prot090811

LB (Luria–Bertani) liquid medium <R>
Lentiviral sgRNA library

Acquire library either by following the associated methodology (see Protocol 1: Single Guide RNA Library Design and Construction [Wang et al. 2016]) or by sourcing library from Addgene (www.addgene.org).

Opti-MEM I Reduced-Serum Medium (Thermo Fisher Scientific 31985-062)
pCMV-dR8.2 packaging plasmid (Addgene 8455)
pCMV-VSV-G pantropic viral envelope plasmid (Addgene 8454)
Penicillin–streptomycin (Sigma-Aldrich P4333)
Phosphate-buffered saline (PBS) <R>
Phusion High-Fidelity PCR Master Mix with HF Buffer (NEB M0531S)
Plasmid Plus Maxi Kit (QIAGEN 12963)
Polybrene (EMD Millipore TR-1003-G)
Puromycin
QIAamp DNA Blood Maxi Kit (QIAGEN 51194)
Sequencing primers for Illumina HiSeq

Read 1 primer:
CGGTGCCACTTTTTCAAGTTGATAACGGACTAGCCTTATTTTAACTTGCTATTTC-
TAGCTCTAAAAC

Indexing primer:
TTTCAAGTTACGGTAAGCATATGATAGTCCATTTTAAAACATAATTTTAAA
ACTGCAAACTACCCAAGAAA

Single guide RNA (sgRNA) barcode polymerase chain reaction (PCR) primers

Forward: AATGATACGGCGACCACCGAGATCTACACCGACTCGGTGCCACTTTT
Reverse: CAAGCAGAAGACGGCATACGAGATCnnnnnTTTCTTGGGTAGTTTGCAGTTTT

The sequence "nnnnnn" denotes a user-specified sample barcode sequence.

Target cells and appropriate cell-culture medium
Trypsin (for adherent cells)
Water (H_2O), PCR grade
X-tremeGENE 9 DNA Transfection Reagent (Roche 06365787001)

Equipment

Access to Illumina HiSeq
Acrodisc syringe filter (0.45-µm; VWR 28144-007)
Bottle top vacuum filter (0.22-µm pore, 150-mL) (Corning 430626)
Centrifuge with rotors for six-well-plate spin infection
Erlenmeyer flask (500-mL)
Gel imager
Heat block (for Step 39)
Luer-Lok Tip syringes (Becton Dickinson)
NanoDrop spectrophotometer (NanoDrop)
Thermocycler
Tissue-culture hood for BL2+ work
Tissue-culture incubator set at 37°C
Tissue-culture-treated plates (six-well, 10- and 15-cm)
Vacuum aspirator
x-tracta gel extractor (USA Scientific 5454-0100)

Cite this protocol as *Cold Spring Harb Protoc*; doi:10.1101/pdb.prot090811

METHOD

We emphasize that you should follow appropriate safety procedures and work in an environment (e.g., BL2+) suitable for handling lentiviruses. A general overview of viral packaging, and issues relating to safety, can be found at https://www.addgene.org/lentiviral/packaging/.

Preparation of Viral Packaging Vector

1. Streak out bacterial stab cultures of pCMV-dR8.2 and pCMV-VSV-G obtained from Addgene on LB-amp plates and incubate overnight at 37°C.
2. Pick a single colony and seed into a 500-mL Erlenmeyer flask containing 100 mL of LB liquid medium containing 100 µg/mL ampicillin.
3. Incubate culture overnight at 37°C.
4. Prepare plasmid DNA from the bacterial culture using the QIAGEN Plasmid Plus Maxi Kit according to the manufacturer's instructions.

Viral Packaging and Titer Test

Day 1

5. Add the following components to make virus production medium (VPM).

DMEM (high glucose, GlutaMAX Supplement)	400 mL
Fetal bovine serum (inactivated)	100 mL
Penicillin (10,000 U/mL)-streptomycin (10 mg/mL)	5 mL

6. Filter medium through a 0.22-µm bottle-top vacuum filter in a tissue-culture hood.
7. Seed 750,000 HEK-293 T cells in a single well of a six-well plate in 2 mL of VPM. Incubate cells overnight at 37°C in a tissue-culture incubator.

Day 2

8. Assemble the following transfection mixture.

Opti-MEM I	50 µL
Lentiviral sgRNA library	1 µg
pCMV-dR8.2	900 ng
pCMV-VSV-G	100 ng
X-tremeGENE 9 DNA Transfection Reagent	5 µL

9. Incubate mixture for 15 min at room temperature and add dropwise to cells to transfect. Incubate cells overnight at 37°C in a tissue-culture incubator.

Day 3

10. Replace medium with 2 mL of VPM. Incubate cells overnight at 37°C in a tissue-culture incubator.

Day 4

11. Harvest viral supernatant from cells and filter through a 0.45-µm Acrodisc syringe filter.

    See Troubleshooting.

12. Set up the following five infections in a six-well tissue-culture-treated plate.

Target cells	5,000,000
Polybrene (10 mg/mL)	2 µL
Filtered virus	0, 125, 250, 500, or 1000 µL
Cell-culture medium	to 2 mL

Some cell lines might not tolerate spin infection and overnight incubation at this density. Adjust cell numbers accordingly for the cell line of interest.

13. Centrifuge plate at 1200*g* for 45 min in a prewarmed centrifuge. After centrifugation, incubate cells overnight at 37°C in a tissue-culture incubator.

Day 5

14. Proceed as follows.
    - For adherent cells, aspirate virus-containing medium, wash cells with PBS, trypsinize cells, and expand each well into a 15-cm tissue-culture-treated plate. Incubate cells overnight at 37°C in a tissue-culture incubator.
    - For suspension lines, pellet cells and aspirate virus-containing medium. Resuspend cells into a 15-cm tissue-culture-treated plate. Incubate cells overnight at 37°C in a tissue-culture incubator.

Day 6

15. Add an appropriate selection dose of puromycin to cells.

    The optimal dose should be determined by performing a puromycin kill curve. The concentration of puromycin that causes death of 100% of unmodified cells in 48–72 h should be used as a selection dose.

Day 9

16. Observe plates. Identify the viral dose required for ~40% cell survival (multiplicity of infection $\asymp$ 0.5), and discard all plates.

Screen Viral Packaging and Infection

Day 1

17. Based on the viral titer test, calculate the volume of virus required to represent the entire library in the cell line of interest 1000-fold (e.g., for a 40,000-sgRNA library, this requires 40,000,000 infected cells—i.e., 100,000,000 total cells, equivalent to 20 times the test infection volume for 5,000,000 cells).
18. Scale-up virus production in 10-cm plates (~10 mL virus produced per plate), seeding 3,750,000 HEK-293 T cells per plate in 10 mL of VPM. Incubate cells overnight at 37°C in a tissue-culture incubator.

Day 2

19. For each plate, assemble the following transfection mixture:

Opti-MEM I	250 µL
Lentiviral sgRNA library	5 µg
pCMV-dR8.2	4.5 µg
pCMV-VSV-G	500 ng
X-tremeGENE 9 DNA Transfection Reagent	25 µL

20. Incubate mixture for 15 min at room temperature and add dropwise to cells to transfect. Incubate cells overnight at 37°C in a tissue-culture incubator.

Day 3

21. Replace medium in plates with 10 mL of fresh VPM. Incubate cells overnight at 37°C in a tissue-culture incubator.

Day 4

22. Harvest viral supernatant from cells and filter through a 0.45-µm Acrodisc syringe filter.

    Viral supernatants can be stored at −80°C for long-term storage, but freeze–thawing will cause a reduction in viral titers (typically ~30%–50% reduction).

23. Calculate the number of wells in a six-well tissue-culture-treated plate required for infection (e.g., for a 40,000 sgRNA library, one needs 40,000,000 infected cells, equivalent to 100,000,000 total cells, thus requiring 20 wells each containing 5,000,000 cells).

 Cite this protocol as *Cold Spring Harb Protoc*; doi:10.1101/pdb.prot090811

24. Assemble a large-scale cell-virus infection mixture according to the following amounts per well:

Target cells	5,000,000
Polybrene (10 mg/mL)	2 µL
Viral dose	giving ~40% cell survival (from Step 16)
Cell culture medium	to 2 mL

Some lines might not tolerate spin infection and overnight incubation at this density—adjust accordingly for your lines of interest.

25. Dispense 2-mL aliquots of the mixture into six-well plates.
26. Centrifuge the plates at 1200*g* for 45 min in a prewarmed centrifuge. After centrifugation, incubate the cells overnight at 37°C in a tissue-culture incubator.

Day 5

27. Proceed as follows.
    - For adherent cells, aspirate virus-containing medium, wash with PBS, trypsinize the cells, and expand each infection into 15-cm tissue-culture-treated plates. Incubate the cells overnight at 37°C in a tissue-culture incubator.
    - For suspension lines, pellet the cells and aspirate virus-containing medium. Resuspend the cells into 15-cm tissue-culture-treated plates. Incubate the cells overnight at 37°C in a tissue-culture incubator.
28. As a control, seed uninfected cells at an identical confluence into a 15-cm tissue-culture-treated plate. Incubate the cells overnight at 37°C in a tissue-culture incubator.

Day 6

29. Add an appropriate selection dose of puromycin to library-infected and uninfected control cells.

    The optimal dose should be determined by performing a puromycin kill curve.

Day 9

30. Observe the plates after 3 d. If cell survival is ≥40% (multiplicity of infection ≈0.5) in the infected population and <5% in the uninfected population, passage the infected cells into fresh medium. Be sure to maintain 1000-fold coverage of the library. With the remaining cells, freeze two pellets for DNA extraction.

    These cells will serve as the initial reference population.

Screen Cell Culture and Library Preparation

After infection and selection of the cell population, all subsequent tissue-culture work should be performed in a BL2 environment.

31. Continue to passage cells, maintaining 1000-fold coverage of the library at each seeding.

    For positive-selection-based screens, the selection agent should be added at 1 wk after infection to allow sufficient time for knockouts to be generated.

32. After 14 population doublings, collect the final cell pellets.
33. Extract genomic DNA from the initial and final cell pellets using the QIAamp DNA Blood Maxi Kit according to the manufacturer's instructions.
34. Calculate the total number of polymerase chain reactions (PCRs) required. Use a 250-fold coverage of the library as input for sgRNA amplification, with 3 µg genomic DNA per 50 µL reaction (e.g., for a 40,000 sgRNA library, requiring 10,000,000 genome equivalents, one needs 66 µg for diploid human cells—i.e., 22 reactions, each with 3 µg of genomic DNA).

35. Use the following per-sample recipe to assemble the total reaction mixture and dispense into PCR strip tubes in 50-µL aliquots on ice.

Genomic DNA	3 µg
Forward sgRNA PCR primer (10 µM)	2 µL
Sample-specific barcoded reverse sgRNA PCR primer (10 µM)	2 µL
Phusion PCR High-Fidelity Master Mix with HF buffer	25 µL
H_2O	to 50 µL

36. Amplify reactions in a thermocycler using the following program.

1 cycle	98°C	2 min
30 cycles	98°C	10 sec
	60°C	15 sec
	72°C	45 sec
1 cycle	72°C	5 min
1 cycle	4°C	hold

37. Pool the reactions and run them on a 1% agarose gel stained with ethidium bromide. Visualize the PCR bands using a standard gel imager.
38. Cut the amplified PCR product using an x-tracta gel extractor tool.
    The expected product size is 274 bp.
39. Extract DNA using the QIAGEN Gel Extraction Kit according to the manufacturer's instructions, eluting in 30 µL of water.
40. Submit extracted PCR products for high-throughput sequencing on an Illumina HiSeq using the custom sequencing primers detailed above (see Reagents).
    A single end run with a 6-bp indexing read should be performed.

Data Analysis

The procedure below describes a simple method for calculating gene scores. A suite of tools (originally designed for analyzing short-hairpin-RNA-based screens) exists for more sophisticated tabulation of gene scores, hit identification, and pathway analysis (Subramanian et al. 2005; Luo et al. 2008; Shao et al. 2013).

41. Process each sample as follows.
    i. Enumerate sgRNA library barcodes using the Bowtie alignment program.
    ii. Add 1 to each sgRNA count.
    iii. Calculate the $\log_2$ fractional abundance of each sgRNA.
42. For each sgRNA of each final sample, subtract the fractional abundance of the sgRNA in the initial sample to determine the $\log_2$ fold change in abundance.
43. To calculate gene scores for each final sample, find the average $\log_2$ fold change of all sgRNAs targeting each gene.
44. To compare between samples, subtract the gene scores between the samples to identify the differentially scoring genes.

TROUBLESHOOTING

Problem (Step 11): The viral titers are too low.
Solution: Low viral production is typically the result of unhealthy HEK-239 T packaging cells. Be sure to check the health of the HEK-239 T cells before and after transfection. Ethanol precipitation of the packaging and transfer vectors can also help eliminate bacterial endotoxin, which strongly inhibits viral production.

Cite this protocol as *Cold Spring Harb Protoc*; doi:10.1101/pdb.prot090811

RECIPES

Ampicillin Stock Solution (100 mg/mL)

Ampicillin (sodium salt [sodium ampicillin], m.w. = 371.40)

Dissolve 1 g of sodium ampicillin in sufficient H_2O to make a final volume of 10 mL. If sterilization is required, prewash a 0.45- or 0.22-µm sterile filter by drawing through 50–100 mL of H_2O. Then pass the ampicillin solution through the washed filter. Store the ampicillin in aliquots at −20°C for 1 yr (or at 4°C for 3 mo).

LB-Ampicillin Agar Plates

Ampicillin, filter-sterilized (10 mg/mL stock)
LB agar

Autoclave 1 L of LB agar. Cool to 55°C. Add 10 mL of ampicillin stock. Pour into Petri dishes (~ 25 mL per 100-mm plate).

LB Agar

Agar (20 g/L)
NaCl (10 g/L; Sigma-Aldrich S9625)
Tryptone (10 g/L; BD 211705)
Yeast extract (5 g/L; BD 212750)

Add H_2O to a final volume of 1 L. Adjust the pH to 7.0 with 5 N NaOH. Autoclave. Pour into Petri dishes (~ 25 mL per 100-mm plate).

LB (Luria-Bertani) Liquid Medium

Reagent	Amount to add
H_2O	950 mL
Tryptone	10 g
NaCl	10 g
Yeast extract	5 g

Combine the reagents and shake until the solutes have dissolved. Adjust the pH to 7.0 with 5 N NaOH (~0.2 mL). Adjust the final volume of the solution to 1 L with H_2O. Sterilize by autoclaving for 20 min at 15 psi (1.05 kg/cm^2) on liquid cycle.

Phosphate-Buffered Saline (PBS)

Reagent	Amount to add (for 1× solution)	Final concentration (1×)	Amount to add (for 10× stock)	Final concentration (10×)
NaCl	8 g	137 mM	80 g	1.37 M
KCl	0.2 g	2.7 mM	2 g	27 mM
Na_2HPO_4	1.44 g	10 mM	14.4 g	100 mM
KH_2PO_4	0.24 g	1.8 mM	2.4 g	18 mM
If necessary, PBS may be supplemented with the following:				
$CaCl_2 \cdot 2H_2O$	0.133 g	1 mM	1.33 g	10 mM
$MgCl_2 \cdot 6H_2O$	0.10 g	0.5 mM	1.0 g	5 mM

PBS can be made as a 1× solution or as a 10× stock. To prepare 1 L of either 1× or 10× PBS, dissolve the reagents listed above in 800 mL of H_2O. Adjust the pH to 7.4 (or 7.2, if required) with HCl, and then add H_2O to 1 L. Dispense the solution into aliquots and sterilize them by autoclaving for 20 min at 15 psi (1.05 kg/cm^2) on liquid cycle or by filter sterilization. Store PBS at room temperature.

REFERENCES

Luo B, Cheung HW, Subramanian A, Sharifnia T, Okamoto M, Yang X, Hinkle G, Boehm JS, Beroukhim R, Weir BA, et al. 2008. Highly parallel identification of essential genes in cancer cells. *Proc Natl Acad Sci* **105:** 20380–20385.

Shao DD, Tsherniak A, Gopal S, Weir BA, Tamayo P, Stransky N, Schumacher SE, Zack TI, Beroukhim R, Garraway LA, et al. 2013. ATARiS: Computational quantification of gene suppression phenotypes from multisample RNAi screens. *Genome Res* **23:** 665–678.

Subramanian A, Tamayo P, Mootha VK, Mukherjee S, Ebert BL, Gillette MA, Paulovich A, Pomeroy SL, Golub TR, Lander ES, et al. 2005. Gene set enrichment analysis: A knowledge-based approach for interpreting genome-wide expression profiles. *Proc Natl Acad Sci* **102:** 15545–15550.

Wang T, Lander ES, Sabatini DM. 2016. Single guide library design and construction. *Cold Spring Harb Protoc* doi:10.1101/pdb.prot090803.

 Cite this protocol as *Cold Spring Harb Protoc*; doi:10.1101/pdb.prot090811

CHAPTER 5

Adeno-Associated Virus–Mediated Delivery of CRISPR–Cas Systems for Genome Engineering in Mammalian Cells

Thomas Gaj[1] and David V. Schaffer[1,2,3,4,5]

[1]*Department of Chemical and Biomolecular Engineering, University of California, Berkeley, California 94720;* [2]*Department of Bioengineering, University of California, Berkeley, California 94720;* [3]*Department of Cell and Molecular Biology, University of California, Berkeley, California 94720;* [4]*Helen Wills Neuroscience Institute, University of California, Berkeley, California 94720*

The CRISPR–Cas9 system has emerged as a highly versatile platform for introducing targeted genome modifications into mammalian cells and model organisms. However, fully capitalizing on the therapeutic potential for this system requires its safe and efficient delivery into relevant cell types. Adeno-associated virus (AAV) vectors are a clinically promising class of engineered gene-delivery vehicles capable of safely infecting a broad range of dividing and nondividing cell types, while also serving as a highly effective donor template for homology-directed repair. Together, CRISPR–Cas9 and AAV technologies have the potential to accelerate both basic research and clinical applications of genome engineering. Here, we present a step-by-step protocol for AAV-mediated delivery of CRISPR–Cas systems into mammalian cells. Procedures are given for the preparation of high-titer virus capable of achieving a diverse range of genetic modifications, including gene knockout and integration.

MATERIALS

It is essential that you consult the appropriate Material Safety Data Sheets and your institution's Environmental Health and Safety Office for proper handling of equipment and hazardous materials used in this protocol.

RECIPES: Please see the end of this protocol for recipes indicated by <R>. Additional recipes can be found online at http://cshprotocols.cshlp.org/site/recipes.

Reagents

AAV (adeno-associated virus) lysis buffer <R>
AflII, BsmBI (New England Biolabs R0580S), and KpnI restriction endonucleases
Agarose gels
Ampicillin
Antibiotic-antimycotic (Thermo Fisher Scientific 15240096)
Benzonase nuclease (Sigma-Aldrich E8263)
DMSO
DNase I (Roche 04716728001)
DNase dilution buffer (10×) <R>
Dry-ice–ethanol bath

[5]Correspondence: schaffer@berkeley.edu

Copyright © Cold Spring Harbor Laboratory Press; all rights reserved
Cite this protocol as *Cold Spring Harb Protoc*; doi:10.1101/pdb.prot086868

Dulbecco's modified Eagle's medium (DMEM)
Escherichia coli TOP10 cells (Thermo Fisher Scientific C4040-03)
Expand High Fidelity PCR System (Roche 11759078001)
Fetal bovine serum (FBS)
Gel extraction kit
Gel-loading dye (10×) <R>
Human embryonic kidney (HEK) 293 T cells (ATCC CRL-1573)
iCycler mix (2×) <R>

This is needed to make the qPCR master mix.

Iodixanol solution (54%) <R>
Iodixanol solutions (15%, 24%, and 40%) <R>
LB solid or liquid medium <R>
Linear polyethylenimine (PEI) (MW 25,000; Polysciences)
$MgCl_2$ (0.15 M)
Oligonucleotides (see Steps 4, 14, 76, and 78)
pAAV–Cas9–sgRNA plasmid

This is available on request to the authors.

PBS (10×) +/− Tween 20 <R>

Include the appropriate concentration of Tween 20 and dilute if necessary. This protocol requires 1× PBS, 1× PBS containing 0.001% Tween 20, and 1× PBS containing 5% Tween 20.

PBS-MK (10×) <R>

Include the appropriate concentration of NaCl and dilute if necessary. This protocol requires 10× PBS-MK containing 1.37 M NaCl, 1× PBS-MK containing 0.137 M NaCl, and 1× PBS-MK containing 2 M NaCl.

pHelper plasmid (available upon request from the authors) (or plasmid that contains adenovirus helper genes)
Polyacrylamide gels
Proteinase K (New England Biolabs P8107S)
Proteinase K incubation buffer (2×) <R>
pXX2 (Cell Biolabs VPK-422) (or plasmid that contains the desired AAV *rep* and *cap* genes)
qPCR master mix <R>
QuickExtract DNA Extraction Solution (Epicentre QE09050)
SURVEYOR Mutation Detection Kit, containing Enhancer, Nuclease, and Stop Solution (Integrated DNA Technologies 706021)
SYBR Safe (Thermo Fisher Scientific S33102)
T4 DNA ligase with buffer (New England Biolabs M0202L)
T4 polynucleotide kinase (New England Biolabs M0201L)
Terrific broth (TB) medium <R>
TBE electrophoresis buffer (10×) <R>

Use diluted in distilled water at 1× strength.

Trypsin-EDTA (Thermo-Fisher Scientific 25300054)

Equipment

96-well flat-bottom tissue-culture plate
Access to DNA sequencing facilities
Benchtop centrifuge
Cell-culture plates (15-cm)
Cell scraper
Conical tubes (sterile, 15- and 50-mL)
Freezer set at −20°C

Cite this protocol as *Cold Spring Harb Protoc*; doi:10.1101/pdb.prot086868

Gel imaging system
Heat block with adjustable temperature
Incubator at 37°C with 5% CO_2
Long blunt-ended cannulas
Microcentrifuge tubes (sterile, 1.7-mL)
Online DNA sequence and analysis tools (see Steps 1–3)
OptiSeal polyallomer centrifuge tubes (4.9-mL capacity; Beckman Coulter 362185)
Plasmid midi or maxiprep kit

An alternative is to use polyethylene glycol (PEG) precipitation (see Step 18).

Plasmid miniprep kit
Polyacrylamide gel electrophoresis (PAGE) apparatus
Preparative ultracentrifuge with fixed-angle rotor
Regular-bevel needle (21-gauge, 1½″)
Ring stand and clamp
Shaking incubator at 37°C
Syringes (sterile, 1- and 3-mL)
Thermocycler
Ultra-15 Centrifugal Filter Units (MWCO 100-kDa; Amicon UFC910024)
Vortex mixer
Water baths set to 37°C and 42°C

METHOD

Cloning

1. Retrieve the DNA sequence of the targeted gene using a reference genome database (e.g., http://www.ncbi.nlm.nih.gov/genome/).
2. Search for potential Cas9 cleavage sites using an online CRISPR design tool (Cradick et al. 2014; Montague et al. 2014) or DNA sequence viewing software.
3. For the *Streptococcus pyogenes* (SpCas9) protein, search the gene sequence for the motif 5′-$G(N)_{19}$-NGG-3′, where 5′-NGG-3′ is the protospacer-adjacent motif (PAM) recognized by SpCas9 (Jinek et al. 2012; Cong et al. 2013; Mali et al. 2013). Alternatively, for the *Staphylococcus aureus* (SaCas9) protein (Ran et al. 2015), search the gene sequence for the motif 5′-$G(N)_{21-24}$-NNGRRT-3′ (where R is A or G).

   A "G" nucleotide is recommended at the 5′ end of the single guide RNA (sgRNA) transcript for efficient expression from the human U6 promoter.

4. Design and order custom sense and antisense oligonucleotides encoding the selected sgRNA protospacer sequences, as shown in Figure 1A.
5. Phosphorylate 1 µM of each oligonucleotide with 5 units (U) of T4 polynucleotide kinase in recommended buffer in a total volume of 20 µL for 30 min at 37°C.
6. Anneal oligonucleotides by incubation for 5 min at 95°C, followed by fast cooling on ice for 5 min.
7. Digest pAAV–Cas9–sgRNA (empty) with BsmBI in recommended buffer for 3 h using 10 U of enzyme per microgram of DNA. Visualize DNA by agarose gel electrophoresis using a 1.2% agarose gel and fluorescent intercalating dye, such as SYBR Safe.

   All AAV vectors and sequences used here are available from the authors on request.

8. Purify the linearized pAAV–Cas9–sgRNA (empty) using a gel extraction kit, according to the manufacturer's instructions.

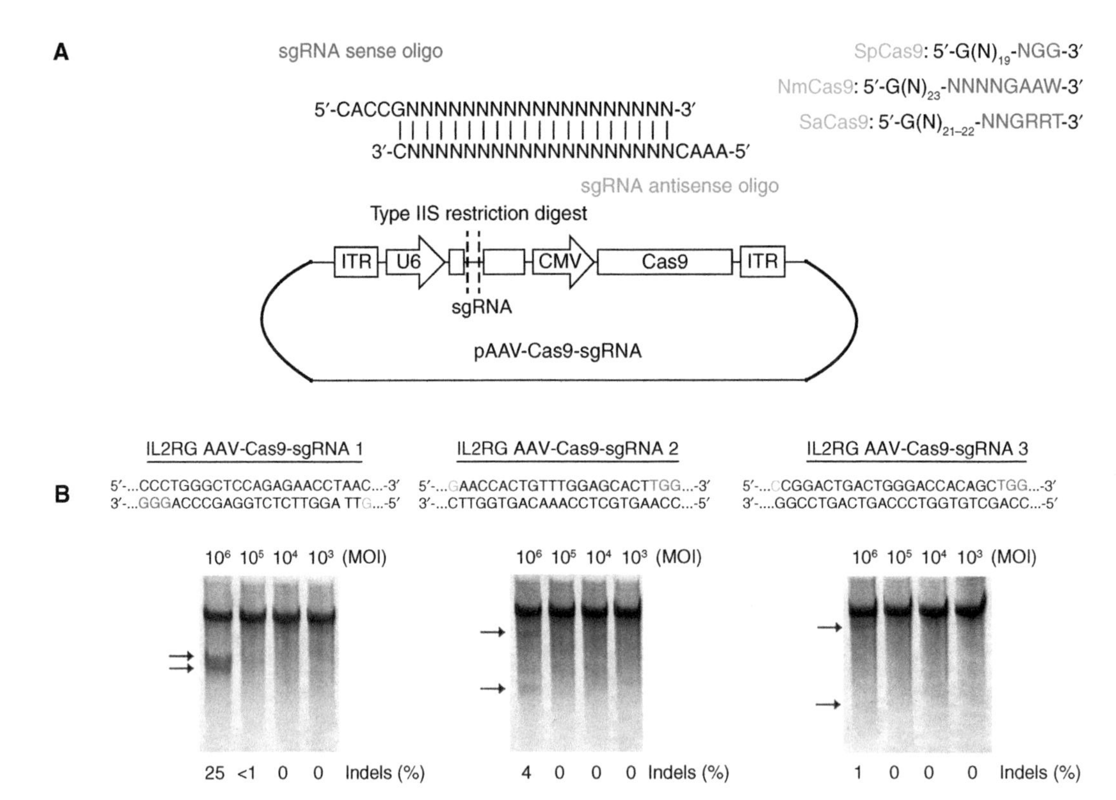

FIGURE 1. Adeno-associated virus (AAV)-mediated delivery of CRISPR–Cas9 for genome editing in mammalian cells. (*A*) Vector preparation. *Streptococcus pyogenes* (Sp), *Neisseria meningitidis* (Nm), and *Staphylococcus aureus* (Sa) Cas9 target sites and sense and antisense oligonucleotides for constructing sgRNA. Sense and antisense sgRNA oligonucleotides encode 5′-CACC-3′ and 5′-AAAC-3′ overhangs, respectively, for insertion into pAAV–Cas9–sgRNA. AAV vectors encoding SpCas9 and NmCas9 should be digested with BsbI, whereas AAV vectors encoding SaCas9 should be digested with BsaI. (*B*) Frequency of endogenous interleukin-2 receptor γ chain (*IL2RG*) gene modification in HEK293 T cells infected with AAV–Cas9–sgRNA of increasing MOI with three different sgRNAs (1–3), as determined by a SURVEYOR nuclease assay. Arrows indicate the position of the expected cleavage product. The protospacer-adjacent motif (PAM) and "G" initiation nucleotide are colored red and blue, respectively. Cas9, CRISPR-associated protein 9; CMV, cytomegalovirus promoter; CRISPR, clustered regularly interspaced short palindromic repeat; indel, insertion or deletion; ITR, inverted terminal repeat; MOI, multiplicity of infection; sgRNA, single guide RNA.

9. Ligate the sgRNA duplex DNA into 20–50 ng of linearized pAAV–Cas9–sgRNA (empty) using 1 U of T4 DNA ligase for 1 h at room temperature.

   A 6:1 molar insert:vector ratio is recommended for ligation.

10. Thaw 100 µL of chemically competent *E. coli* TOP10 cells on ice and mix gently with ligated pAAV–Cas9–sgRNA.
11. Keep the cells on ice for 30 min. Heat-shock the mixture for 35 sec at 42°C and recover the cells in 1 mL of lysogeny broth (LB) for at least 30 min at 37°C with shaking.
12. Spread 50–100 µL of bacterial cell culture on a LB agar plate with 100 µg/mL ampicillin and incubate overnight at 37°C.
13. The following day, inoculate 2–4 mL of terrific broth (TB) medium containing 100 µg/mL ampicillin with one colony from the LB agar plate and culture overnight at 37°C with shaking.
14. Purify pAAV–Cas9–sgRNA by plasmid miniprep and confirm plasmid identity by DNA sequencing using the primer U6 Seq (5′-GACTGTAAACACAAAGATATTAGTAC-3′).
15. Test the ability of Cas9 to induce modifications at the genomic target site in mammalian cell culture by transient transfection of pAAV–Cas9–sgRNA using the procedures described in Steps 60–75.

   It is strongly recommended that this control be performed.

 Cite this protocol as *Cold Spring Harb Protoc*; doi:10.1101/pdb.prot086868

16. Thaw 100 μL of chemically competent *E. coli* TOP10 cells on ice and mix gently with 100 ng of pAAV–Cas9–sgRNA plasmid. Transform as above (Steps 10–12).
17. The following day, inoculate 50–100 mL of TB medium containing 100 μg/mL ampicillin with one colony and grow overnight at 37°C with shaking.
18. Purify plasmid DNA by plasmid midiprep or maxiprep, according to the manufacturer's instructions, or by PEG precipitation.
19. Store plasmid at −20°C until transfection.

Adeno-Associated Virus Production

20. Maintain HEK293 T cells in DMEM containing 10% (v/v) FBS and 1% antibiotic-antimycotic at 37°C in a fully humidified atmosphere with 5% CO_2.
21. Seed HEK293 T cells onto a 15-cm plate at a density of 2.5–3 × 10^7 cells/plate.
22. At 24 h after seeding, or once cells are ~90% confluent, add 15 μg pAAV–Cas9–sgRNA, 15 μg pXX2, and 15 μg pHelper plasmids to 4 mL of cell-culture medium in a sterile 15-mL conical tube.
23. Add 135 μL PEI (1 μg/μL) and mix immediately by vortexing for 10 sec.

    The volume of PEI is based on a 3 to 1 ratio of PEI (μg) to total DNA (μg).
24. Incubate transfection solution for 10 min at room temperature.
25. Add transfection solution dropwise to cells.
26. (Optional) Change media 8–12 h posttransfection to reduce transfection-reagent-associated toxicity.
27. Harvest virus from cells 48–72 h posttransfection by manually dissociating cells from plate using a cell scraper and pipetting media and cells into 50-mL conical tubes.
28. Pellet cells by centrifugation at 1500*g* for 5 min at room temperature.
29. Remove media and resuspend cells in 2 mL of lysis buffer for each 15-cm plate.
30. Freeze–thaw cells three times using a dry-ice–ethanol bath and a 37°C water bath.

    The cell lysate can be stored at −20°C after the third freeze.
31. Incubate the cells with 10 U of benzonase per milliliter of cell lysate. Incubate the samples for 30 min at 37°C.
32. Centrifuge cell lysate at 10,000*g* for 10 min at room temperature.
33. Transfer supernatant to new tubes and store at 4°C until purification.

Iodixanol Density Gradient Centrifugation

34. Pipette 1.2 mL of 15% iodixanol solution into an OptiSeal polyallomer centrifuge tube.

    Ensure that each iodixanol solution contains the appropriate amount of 10× PBS-MK containing 1.37 M NaCl, 1× PBS-MK containing 0.137 M NaCl, or 1× PBS-MK containing 2 M NaCl.
35. Underlay the 15% iodixanol solution with 0.7 mL of 24% iodixanol solution containing phenol red using a long blunt-ended cannula attached to a 3-mL syringe.
36. Underlay the 24% iodixanol solution with 0.6 mL of 40% iodixanol solution.
37. Underlay the 40% iodixanol solution with 0.6 mL of 54% iodixanol solution containing phenol red.
38. Gently pipette 1.8 mL of crude lysate onto each gradient.
39. Weigh tubes to ensure that they are properly balanced. Use AAV lysis buffer to balance tubes as necessary and seal tubes using the caps provided.

40. Set preparative ultracentrifuge to slow acceleration and deceleration settings and centrifuge gradients at 174,000*g* for 2 h at 18°C.
41. Carefully remove centrifuge tubes from the rotor and secure the centrifuge tube in a clamp attached to a ring stand. Remove the cap.
42. Carefully puncture the tube at the interface between the 40% and 50% iodixanol solutions using a 21-gauge 1½″ regular-bevel needle attached to a 1-mL syringe.
43. Collect the bottom four-fifths of the 40% iodixanol solution (bevel up) and the top one-fifth of the 54% iodixanol solution (bevel down) (Zolotukhin et al. 1999).

    Contaminating proteins from the cell lysate will be present in a band at the interface between the 24% and 40% iodixanol layers. Do not collect the protein band.
44. Store collected fractions in a sterile 1.7-mL microcentrifuge tube or a 15-mL conical tube at 4°C until further purification.

Buffer Exchange and Concentration

45. Incubate Ultra-15 Centrifugal Filter Unit in 1× PBS containing 5% Tween 20 for 30 min at room temperature. After incubation, wash filter once with 1× PBS containing 0.001% Tween 20.
46. Dilute collected iodixanol fraction to 15 mL in 1× PBS containing 0.001% Tween 20 and add to Ultra-15 Centrifugal Filter Unit.
47. Centrifuge at 4000*g* for 30 min or until solution has been concentrated to <2 mL.
48. Add 15 mL of 1× PBS containing 0.001% Tween-20 and mix well.
49. Repeat Steps 47 and 48 three times or until all iodixanol has been eliminated and the viscosity of the solution is similar to that of 1× PBS containing 0.001% Tween 20.
50. Concentrate the virus to the desired volume and store at 4°C.

Viral Titering

51. Combine 1 µL of virus with 5 µL of 10× DNase dilution buffer, 0.5 µL of DNase I, and 43.5 µL of water. Incubate virus sample for 30 min at 37°C.
52. Incubate the sample for 10 min at 75°C to inactivate DNase I.
53. Add 60 µL of 2× proteinase K incubation buffer and 10 µL of proteinase K to virus sample and incubate for at least 1 h at 37°C.
54. Incubate for 20 min at 95°C to inactivate the proteinase K.
55. Create 10-fold serial dilutions of pAAV–SpCas9–sgRNA plasmid between 0.2 ng/µL and 0.02 pg/µL for generating a standard curve.
56. Prepare qPCR master mix containing iCycler mix.
57. Dilute virus sample 10-fold for qPCR.
58. Combine 15 µL of qPCR master mix with 5 µL of virus sample or linear plasmid for the standard curve and run qPCR using the following protocol.

1 cycle	5 min	95°C
40 cycles	30 sec	95°C
	30 sec	60°C
	20 sec	72°C

59. Plot threshold cycle (C_t) values for standards against the $\log_{10}$ of the starting plasmid copy number. Correlate the C_t value of the virus sample to the copy number of the standard from a corresponding C_t value.

    See Troubleshooting.

Cite this protocol as *Cold Spring Harb Protoc*; doi:10.1101/pdb.prot086868

Genome Modification

60. Seed HEK293 T cells (or the most relevant cell type) onto a 96-well flat-bottom tissue-culture plate at a density of 4×10^4 cells per well.
61. At 24 h after seeding, add AAV–SpCas9–sgRNA vector to cells at a genomic multiplicity of infection (MOI) of ~10^6.

    Vector can be diluted in serum-containing medium.
62. (Recommended) Vary the MOI five- to 10-fold from 10^6 to 10^2 to further assess vector activity.
63. At 72 h after infection, wash cells once with 1× PBS and isolate infected cells by trypsin–EDTA digestion.

    See Troubleshooting.
64. Use benchtop centrifuge to collect cells (1100*g* for 3 min).
65. Remove supernatant and resuspend cells by vigorous pipetting with 50 µL of QuickExtract DNA Extraction Solution.
66. Incubate samples for 15 min at 65°C, followed for 15 min at 98°C. Hold samples at 4°C or store indefinitely at −80°C.
67. Amplify the targeted genomic region by polymerase chain reaction (PCR) using the Expand High Fidelity PCR System. Perform a 50-µL PCR.

Template DNA	3 µL
Expand high fidelity buffer (10×) with $MgCl_2$	5 µL
Each primer	0.4 µM
High fidelity *Taq* DNA polymerase	0.5 µL
DMSO	5%
Water	to 50 µL

68. Verify amplification by agarose gel electrophoresis.
69. Denature and re-anneal the PCR amplicon to generate mismatched duplex DNA for the SURVEYOR nuclease assay using the following profile.

95°C	10 min
95°C–85°C	−2°C/sec
85°C–25°C	−0.1°C/sec
4°C	Hold

70. Mix 10 µL of heteroduplex DNA with 1 µL of 0.15 M $MgCl_2$, 1 µL of SURVEYOR Enhancer S, and 1 µL of SURVEYOR Nuclease S. Incubate the reaction for 1 h at 42°C.
71. Quench reaction with 1 µL of Stop solution and add 2 µL of 10× DNA gel loading dye to each sample.
72. Load the samples in a PAGE apparatus on a 10%–14% TBE acrylamide gel and run the gel at 140–180 V until the xylene cyanol band from the gel loading dye is located in the middle or bottom third of the gel.
73. Remove the gel and stain with 10 µL of SYBR Safe in 30 mL of 1× TBE electrophoresis buffer for 10 min. Wash the gel at least once with water.
74. Visualize the gel using a gel imaging system and measure the density or intensity of each band.
75. Determine the percentage gene modification by measuring the fraction of parental band cleaved at the anticipated location, as described previously (Guschin et al. 2010).

    Representative results are shown in Figure 1B. See Troubleshooting.

Gene Targeting

Nuclease-induced double-strand breaks can stimulate integration of donor DNA into an endogenous locus through homology-directed repair (HDR) (Rouet et al. 1994; Bibikova et al. 2001). AAV, in particular, can enhance gene targeting by >1000-fold compared with plasmid DNA (Russell and Hirata 1998; Jang et al. 2011; Asuri et al. 2012; Gaj et al. 2015).

76. To construct AAV vectors for gene targeting, design primers to PCR-amplify "left" and "right" homology arms that flank the intended modified sequence. (The 5′ [sense] primer for the "left" homology arm should encode an AflII restriction site, and the 3′ [antisense] primer for the "right" homology arm should encode a KpnI restriction site.) Situate the Cas9 cleavage site within 50 bp of each homology arm.

    Optimal homology arm length ranges from 0.5 to 1.5 kb.

77. Encode single-base modifications on the 3′ and 5′ ends of the antisense and sense primers of the "left" and "right" homology arms, respectively. Encode a silent restriction site on the donor template for downstream analysis.
78. For gene integration, design a second set of primers to amplify the gene of interest (GOI), with 20–30 nucleotides of sequence complementary to the 3′ and 5′ ends of the "left" and "right" homology arms, respectively. Ensure that the Cas9 cleavage site is not present in the donor construct.

    Because of the carrying capacity of AAV, the donor template should not exceed 4.7 kb.

79. PCR-amplify the homology arms from genomic DNA and the GOI from cDNA or plasmid DNA using the primers designed in Steps 76 and 78.
80. Gel-purify the homology-arm-encoding and GOI-encoding amplicons using a gel extraction kit, according to the manufacturer's instructions.
81. Fuse the amplicons together by overlap PCR or Gibson assembly (Gibson et al. 2009) to generate the donor template.
82. Gel-purify the donor template by gel extraction and digest both it and the AAV plasmid (e.g., AAV–Cas9–sgRNA) with AflII and KpnI restriction enzymes.
83. Ligate the donor template into 20–50 ng of digested AAV plasmid and transform into cells, as described in Steps 21–26.
84. Purify the AAV donor plasmid by miniprep and confirm plasmid identity by DNA sequencing.
85. Construct the accompanying AAV–Cas9–sgRNA vector, and package and purify both it and the AAV donor, as described in Steps 20–59.
86. Infect cells with purified AAV vectors, as described in Steps 60–62.

    Cells infected with AAV donor vector containing a selectable marker, such as a puromycin-resistance gene or enhanced green fluorescent protein (EGFP), can be subjected to antibiotic selection or harvested for fluorescence-activated cell sorting (FACS) at 72 h after infection. Limiting dilution is recommended for isolation and expansion of clonal cell lines.

87. Isolate genomic DNA, as described in Steps 63–66, and PCR-amplify the genomic target across the integration junctions using the Expand High Fidelity PCR System. If the donor template contained a silent restriction site, evaluate the integration frequency by restriction digestion analysis. Determine gene modification by measuring the fraction of the parental band cleaved at the anticipated location. Finally, use DNA sequencing to confirm the presence of gene modifications.

    See Troubleshooting.

TROUBLESHOOTING

Problem (Step 59): The calculated titer value is too low.

Solution: A low titer can arise for a number of reasons, including using an impure plasmid preparation, mutations within the AAV plasmid from modification of inverted terminal repeats, toxicity from PEI transfection, off-target cleavage within the AAV vector genome by Cas9, or the vector being released from cells. Possible solutions include using phenol–chloroform extraction of the

AAV plasmid to improve vector purity, diagnostic restriction digestion of the AAV vector to establish vector integrity, transfection by calcium phosphate to eliminate the possibility of PEI-induced toxicity, and harvesting the cells 48 h after transfection.

Problem (Step 63): Infectivity is low.

Solution: Because of differences in primary receptor usage and capsid composition, many naturally occurring AAV vectors display differential infection abilities both in vitro and in vivo. Therefore, the use of AAV vectors with the intended cell or tissue tropism is essential for efficient gene delivery. Infection by AAV serotypes can be measured by using a fluorescent reporter gene, such as that encoding EGFP. Engineered or evolved AAV vectors with improved or altered tropism can also be used to enhance infection (Kotterman and Schaffer 2014).

Problem (Step 75): The efficiency of genome modification is low.

Solution: Poor genome modification could be due to a number of factors, including low levels of Cas9-mediated cleavage at the genomic target site, terminal truncations within the AAV vector genome, and low levels of Cas9 expression. Test the ability of the Cas9–sgRNA complex to induce modifications at the genomic target by transient transfection. Because of the limited carrying capacity of AAV, packaging a single vector containing both a large Cas9 variant (such as SpCas9) and sgRNA could lead to vectors with truncations at the 5′ end of the vector genome (Senis et al. 2014). Use Southern blot analysis to establish whether truncations are present. To minimize vector genome heterogeneity, SpCas9 and sgRNA can be delivered using two separate particles (Swiech et al. 2015). Smaller Cas9 orthologs, such as *Neisseria meningitidis* (NmCas9) (Hou et al. 2013) and SaCas9 (Ran et al. 2015), can also be used to induce genome modifications from a single AAV particle despite their more restrictive PAM requirements. Finally, confirm that the promoter is providing high levels of expression in the desired cell type by western blotting or by the use of a fluorescent reporter gene, such as one encoding EGFP.

Problem (Step 87): The efficiency of integration is low.

Solution: No integration could be the result of insufficient homology arm length, low levels of Cas9 activity, or poor infectivity. Test the ability of the Cas9–sgRNA complex to induce modifications at the genomic target by transient transfection. Use an alternative sgRNA if activity is low. In addition, test the ability of the AAV donor vector in combination with Cas9 to mediate HDR by transient transfection. Modify homology arm length in cases where the existing donor template does not trigger integration. Use of small molecules that inhibit nonhomologous end joining (NHEJ) can also enhance HDR (Chu et al. 2015; Wurst et al. 2015; Yu et al. 2015).

RECIPES

AAV Lysis Buffer

Reagent	Final concentration
Tris-HCl (pH 8.0)	50 mM
NaCl	150 mM

Sterilize using a disposable 0.22-µm vacuum filtration system in a tissue-culture hood and store at room temperature.

DNase Dilution Buffer (10×)

Reagent	Final concentration
Tris-HCl (pH 7.4)	250 mM
$MgCl_2$	100 mM

Sterilize using a disposable 0.22-µm vacuum filtration system in a tissue-culture hood and store at room temperature.

Gel-Loading Dye (10×)

Reagent	Final concentration
Glycerol	60%
EDTA	0.2 M
Bromophenol blue	0.5%
Xylene cyanol	0.5%

Adjust to pH 8.0. Store at room temperature.

iCycler Mix (2×)

Reagent	Volume/final concentration (for 1 mL)
PCR buffer (10×; 200 mM Tris-HCl, 500 mM KCl, pH 8.4)	200 µL
$MgCl_2$	25 mM
dNTPs	10 mM

Sterilize using a disposable 0.22-µm vacuum filtration system in a tissue-culture hood and store at room temperature.

Iodixanol Solution (54%)

Reagent	Volume to add
OptiPrep Density Gradient Medium (60% iodixanol; Sigma-Aldrich D1556)	40 mL
PBS-MK (10×, containing 1.37 M NaCl) <R>	4.44 mL

Sterilize the 54% iodixanol solution using a disposable 0.22-µm vacuum filtration system in a tissue-culture hood and store at room temperature. Add 60 µL of 0.5% phenol red to a 12-mL aliquot of the 54% iodixanol solution.

Iodixanol Solutions (15%, 24%, and 40%)

Reagent	Volume (for 15%)	Volume (for 24%)	Volume (for 40%)
Iodixanol solution (54%) <R>	4.72 mL	5.56 mL	7.41 mL
PBS-MK (1×, containing 2 M NaCl) <R>	8.50 mL	–	–
PBS-MK (1×, containing 0.137 M NaCl) <R>	3.78 mL	6.44 mL	2.59 mL
Phenol red (0.5%)	–	60 µL	–
Total	17 mL	~ 12 mL	10 mL

Sterilize each solution using a disposable 0.22-µm vacuum filtration system in a tissue-culture hood and store at room temperature.

LB Solid or Liquid Medium

10 g	Tryptone
5 g	Yeast extract
5 g	NaCl
20 g	Agar (for solid medium only; omit for liquid medium)

Combine the ingredients and bring to 1 L with ddH_2O. Autoclave and store at room temperature or 4°C.

 Cite this protocol as *Cold Spring Harb Protoc*; doi:10.1101/pdb.prot086868

PBS (10×) +/− Tween 20

Reagent	Final concentration (10×)
Na_2HPO_4 (pH 7.4)	100 mM
KH_2PO_4	18 mM
NaCl	1.37 M
KCl	27 mM
Tween 20	as required

Sterilize 10× PBS without Tween 20 using a disposable 0.22-µm vacuum filtration system in a tissue-culture hood and store at room temperature. Dilute to 1× and/or add Tween 20 as needed: For 1× PBS containing 0.001% Tween 20, dilute 10× PBS 1:9 in distilled water to give a volume of 500 mL, and add 5 µL of Tween 20. For 1× PBS containing 5% Tween 20, dilute 10× PBS 1:9 in distilled water to give a volume of 475 mL, and then add 25 mL of Tween 20 for a final volume of 0.5 L.

PBS-MK (10×)

Reagent	Final concentration (10×)
Na_2HPO_4 (pH 7.4)	100 mM
KH_2PO_4	18 mM
NaCl	as appropriate
$MgCl_2$	10 mM
KCl	25 mM

Sterilize 10× PBS-MK containing the appropriate concentration of NaCl using a disposable 0.22-µm vacuum filtration system in a tissue-culture hood and store at room temperature. When required, dilute to 1× in distilled water.

Proteinase K Incubation Buffer (2×)

Reagent	Final concentration
Tris-HCl (pH 8.0)	10 mM
Na_2EDTA	20 mM
$NaCl_2$	20 mM

Sterilize using a disposable 0.22-µm vacuum filtration system in a tissue-culture hood and store at room temperature.

qPCR Master Mix

Reagent	Volume (for 150 µL)
iCycler mix (2×) <R>	100 µL
Sense primer (1 µM)	2 µL
Antisense primer (1 µM)	2 µL
Fluorescein (1 µM)	2 µL
SYBR Green (40×) (Thermo Fisher Scientific S-7563)	2 µL
Taq DNA polymerase:JumpStart *Taq* Antibody (1:1) (New England Biolabs M0267X)	2 µL
Distilled water	40 µL

Sterilize using a disposable 0.22-µm vacuum filtration system in a tissue-culture hood and store at room temperature.

TBE Electrophoresis Buffer (10×)

Reagent	Quantity (for 1 L)	Final concentration
Tris base	121.1 g	1 M
Boric acid	61.8 g	1 M
EDTA (disodium salt)	7.4 g	0.02 M

Prepare with RNase-free H_2O. Dilute 100 mL to 1 L to make gel running **buffer**. Store for up to 6 mo at room temperature.

Terrific Broth (TB) Medium

Reagent	Quantity	Final concentration
Yeast extract	24 g	24 g/L
Tryptone	20 g	20 g/L
Glycerol	4 mL	4 mL/L
Phosphate buffer (0.17 M KH_2PO_4, 0.72 M K_2HPO_4)	100 mL	0.017 M KH_2PO_4 and 0.072 M K_2HPO_4

Add 900 mL of deionized water to 24 g of yeast extract, 20 g of tryptone, and 4 mL of glycerol. Shake or stir until the solutes have dissolved and sterilize by autoclaving for 20 min at 15 psi (1.05 kg/cm^2). Allow the solution to cool to ~60°C and add 100 mL of sterile phosphate buffer. Store TB at room temperature; it will keep for at least 1 yr.

ACKNOWLEDGMENTS

This work was supported by the National Institutes of Health Grant R01EY022975.

REFERENCES

Asuri P, Bartel MA, Vazin T, Jang JH, Wong TB, Schaffer DV. 2012. Directed evolution of adeno-associated virus for enhanced gene delivery and gene targeting in human pluripotent stem cells. *Mol Ther* **20:** 329–338.

Bibikova M, Carroll D, Segal DJ, Trautman JK, Smith J, Kim YG, Chandrasegaran S. 2001. Stimulation of homologous recombination through targeted cleavage by chimeric nucleases. *Mol Cell Biol* **21:** 289–297.

Chu VT, Weber T, Wefers B. 2015. Increasing the efficiency of homology-directed repair for CRISPR–Cas9-induced precise gene editing in mammalian cells. *Nat Biotechnol* **33:** 543–548.

Cong L, Ran FA, Cox D, Lin S, Barretto R, Habib N, Hsu PD, Wu X, Jiang W, Marraffini LA, et al. 2013. Multiplex genome engineering using CRISPR/Cas systems. *Science* **339:** 819–823.

Cradick TJ, Qiu P, Lee CM, Fine EJ, Bao G. 2014. COSMID: A web-based tool for identifying and validating CRISPR/Cas off-target sites. *Mol Ther Nucleic Acids* **3:** e214.

Gaj T, Epstein BE, Schaffer DV. 2015. Genome engineering using adeno-associated virus: Basic and clinical research applications. *Mol Ther* doi: 10.1038/mt.2015.151.

Gibson DG, Young L, Chuang RY, Venter JC, Hutchison CA III, Smith HO. 2009. Enzymatic assembly of DNA molecules up to several hundred kilobases. *Nat Methods* **6:** 343–345.

Guschin DY, Waite AJ, Katibah GE, Miller JC, Holmes MC, Rebar EJ. 2010. A rapid and general assay for monitoring endogenous gene modification. *Methods Mol Biol* **649:** 247–256.

Hou Z, Zhang Y, Propson NE, Howden SE, Chu LF, Sontheimer EJ, Thomson JA. 2013. Efficient genome engineering in human pluripotent stem cells using Cas9 from *Neisseria meningitidis*. *Proc Natl Acad Sci* **110:** 15644–15649.

Jang JH, Koerber JT, Kim JS, Asuri P, Vazin T, Bartel M, Keung A, Kwon I, Park KI, Schaffer DV. 2011. An evolved adeno-associated viral variant enhances gene delivery and gene targeting in neural stem cells. *Mol Ther* **19:** 667–675.

Jinek M, Chylinski K, Fonfara I, Hauer M, Doudna JA, Charpentier E. 2012. A programmable dual-RNA-guided DNA endonuclease in adaptive bacterial immunity. *Science* **337:** 816–821.

Kotterman MA, Schaffer DV. 2014. Engineering adeno-associated viruses for clinical gene therapy. *Nat Rev Genet* **15:** 445–451.

Mali P, Yang L, Esvelt KM, Aach J, Guell M, DiCarlo JE, Norville JE, Church GM. 2013. RNA-guided human genome engineering via Cas9. *Science* **339:** 823–826.

Montague TG, Cruz JM, Gagnon JA, Church GM, Valen E. 2014. CHOPCHOP: A CRISPR/Cas9 and TALEN web tool for genome editing. *Nucleic Acids Res* **42:** W401–W407.

Ran FA, Cong L, Yan WX, Scott DA, Gootenberg JS, Kriz AJ, Zetsche B, Shalem O, Wu X, Makarova KS, et al. 2015. *In vivo* genome editing using *Staphylococcus aureus* Cas9. *Nature* **520:** 186–191.

Rouet P, Smih F, Jasin M. 1994. Expression of a site-specific endonuclease stimulates homologous recombination in mammalian cells. *Proc Natl Acad Sci* **91:** 6064–6068.

Russell DW, Hirata RK. 1998. Human gene targeting by viral vectors. *Nat Genet* **18:** 325–330.

Senis E, Fatouros C, Grosse S, Wiedtke E, Niopek D, Mueller AK, Borner K, Grimm D. 2014. CRISPR/Cas9-mediated genome engineering: An adeno-associated viral (AAV) vector toolbox. *Biotechnol J* **9:** 1402–1412.

Swiech L, Heidenreich M, Banerjee A, Habib N, Li Y, Trombetta J, Sur M, Zhang F. 2015. *In vivo* interrogation of gene function in the mammalian brain using CRISPR–Cas9. *Nat Biotechnol* **33:** 102–106.

Wurst W, Sander S, Rajewsky K, Kuhn R, Maruyama T, Dougan SK, Truttmann MC, Bilate AM, Ingram JR, Ploegh HL. 2015. Increasing the efficiency of precise genome editing with CRISPR–Cas9 by inhibition of nonhomologous end joining. *Nat Biotechnol* **33:** 538–542.

Yu C, Liu Y, Ma T, Liu K, Xu S, Zhang Y, Liu H, La Russa M, Xie M, Ding S, et al. 2015. Small molecules enhance CRISPR genome editing in pluripotent stem cells. *Cell Stem Cell* **16:** 142–147.

Zolotukhin S, Byrne BJ, Mason E, Zolotukhin I, Potter M, Chesnut K, Summerford C, Samulski RJ, Muzyczka N. 1999. Recombinant adeno-associated virus purification using novel methods improves infectious titer and yield. *Gene Ther* **6:** 973–985.

Cite this protocol as *Cold Spring Harb Protoc*; doi:10.1101/pdb.prot086868

CHAPTER 6

Detecting Single-Nucleotide Substitutions Induced by Genome Editing

Yuichiro Miyaoka,[1] Amanda H. Chan,[1] and Bruce R. Conklin[1,2,3]

[1]*Gladstone Institute of Cardiovascular Disease, San Francisco, California 94158;* [2]*Departments of Medicine, and Cellular and Molecular Pharmacology, University of California, San Francisco, California 94143*

The detection of genome editing is critical in evaluating genome-editing tools or conditions, but it is not an easy task to detect genome-editing events—especially single-nucleotide substitutions—without a surrogate marker. Here we introduce a procedure that significantly contributes to the advancement of genome-editing technologies. It uses droplet digital polymerase chain reaction (ddPCR) and allele-specific hydrolysis probes to detect single-nucleotide substitutions generated by genome editing (via homology-directed repair, or HDR). HDR events that introduce substitutions using donor DNA are generally infrequent, even with genome-editing tools, and the outcome is only one base pair difference in 3 billion base pairs of the human genome. This task is particularly difficult in induced pluripotent stem (iPS) cells, in which editing events can be very rare. Therefore, the technological advances described here have implications for therapeutic genome editing and experimental approaches to disease modeling with iPS cells.

THE CHALLENGE OF DETECTING HOMOLOGY-DIRECTED REPAIR EVENTS IN iPS CELLS

The discovery of human induced pluripotent stem (iPS) cells (Takahashi et al. 2007; Yu et al. 2007) provided a great opportunity to study human disease phenotypes in vitro. However, iPS-cell disease modeling was limited to observational studies, because iPS cells were resistant to the homologous recombination methods used for mouse embryonic stem cells. Site-specific nucleases (Gaj et al. 2013) have revolutionized our ability to make isogenic iPS-cell models that exactly reflect patients' pathological mutations, or to precisely revert pathological mutations to a healthy sequence. Many human diseases are caused by a single point mutation, so it is critical to make accurate disease models with as few additional changes as possible. Traditional gene targeting methods to generate mutant cell lines use antibiotic-resistance markers, leaving a genetic "scar" that can interfere with studying the resulting phenotype (da Cunha Santos et al. 2011; Moore et al. 2012). A major challenge of nuclease-driven precise mutagenesis in iPS cells is that the mutations can be <1% of the cells (Chen et al. 2011; Soldner et al. 2011; Ding et al. 2013), and isolating the right recombinant iPS clones is difficult without antibiotic selection. Furthermore, using high levels of nuclease or increasing nuclease activity runs the risk of reducing the fidelity of mutagenesis (Gupta et al. 2011; Pattanayak et al. 2011; Fu et al. 2013; Hsu et al. 2013).

As multiple groups are now planning to use genome editing to correct point mutations for therapeutic purposes, it becomes even more critical to develop methods that accurately measure homology-directed repair (HDR) events with a wide dynamic range. Initial studies (Chen et al.

[3]Correspondence: bconklin@gladstone.ucsf.edu

Copyright © Cold Spring Harbor Laboratory Press; all rights reserved

Cite this introduction as *Cold Spring Harb Protoc*; doi:10.1101/pdb.top090845

2011; Soldner et al. 2011; Ding et al. 2013) suggest that the highest-fidelity editing conditions would favor using lower concentrations of nucleases or replacing them with nickases to avoid off-target mutations (Mali et al. 2013; Ran et al. 2013). However, using nickases to improve fidelity also lowers efficiency so that mutants are increasingly rare. Therefore, genome engineering is faced with a logistical challenge: Precise mutagenesis with the highest fidelity results in rarer mutagenic events, but isolating a rare mutant cell without antibiotic resistance amid the hundreds of otherwise identical cells is exceedingly difficult.

DROPLET DIGITAL POLYMERASE CHAIN REACTION FOR DETECTING HDR EVENTS

In the accompanying protocol, we provide a method that uses droplet digital polymerase chain reaction (ddPCR) (Hindson et al. 2011) to efficiently and quantitatively detect rare single-base mutations for precise genome editing (see Protocol 1: Using Digital Polymerase Chain Reaction to Detect Single-Nucleotide Substitutions Induced by Genome Editing [Miyaoka et al. 2016]). ddPCR detects, in a robust manner, single-nucleotide substitutions that are introduced by HDR, even in cell types that are difficult to engineer, such as iPS cells. It partitions a reaction into more than 10,000 nanoliter-scale water-in-oil droplets, each of which contains only a few copies of the genome. In this way, ddPCR detects rare genome-editing events.

Because of its high sensitivity and quantitative performance, this method can be used to validate genome-editing tools, isolate cell lines with single-nucleotide substitutions, and perform genotyping (Miyaoka et al. 2014). Substitution of single nucleotides in human cells, especially human pluripotent stem cells, is a highly valuable approach for disease modeling and future cell therapies. This method will empower scientists to fully exploit the advantages of genome-editing tools.

THE FUTURE OF GENOME EDITING IN iPS CELLS

Perhaps the most immediate impact of genome editing in iPS cells is in disease modeling, because it provides a clear path to experiments with isogenic controls. The value of isogenic controls is well known to researchers who work in model organisms, such as mice, in which inbreed strains are considered essential for well-controlled experiments. Human pluripotent stem cells were so difficult to engineer that only eight gene-targeting events were reported within 10 years after the first isolation of human embryonic stem cells (Giudice and Trounson 2008). After the discovery of human iPS cells (Takahashi et al. 2007; Yu et al. 2007), the difficulties with genome editing limited disease modeling in iPS cells to observations without isogenic controls. The advent of transcription-activator-like effector nucleases (TALENs), the CRISPR–Cas9 system, and the methods we describe here has dramatically changed disease modeling with iPS cells (Fig. 1A).

Our own initial studies with iPS cells from patients with a disease, long QT syndrome (Spencer et al. 2014), made us aware that using nonisogenic controls impeded our efforts to develop assays that bring out the disease phenotype (Fig. 1A). We found that each patient iPS-cell line can have unique physiological features that are driven by their genetic background, making it harder to determine the physiological effects that were solely due to the disease mutation. With the advent of genome editing, we can make putative disease mutations in a reference cell line so that assay development can proceed in an isogenic background. Once the assays have been optimized, it is much easier to examine the more challenging patient-derived lines (Fig. 1A). In our experience, making the isogenic iPS lines from a robust reference cell line dramatically decreases the time and cost of a study.

Precise genome editing in iPS cells also has ramifications for how the community approaches large-scale collections of iPS cells harboring genetic diseases. In the past, the only route to making these iPS-cell collections was to collect large numbers of patient samples (Fig. 1B). In fact, multiple international efforts have been launched to build large collections of diverse iPS-cell lines, with more

Cite this introduction as *Cold Spring Harb Protoc*; doi:10.1101/pdb.top090845

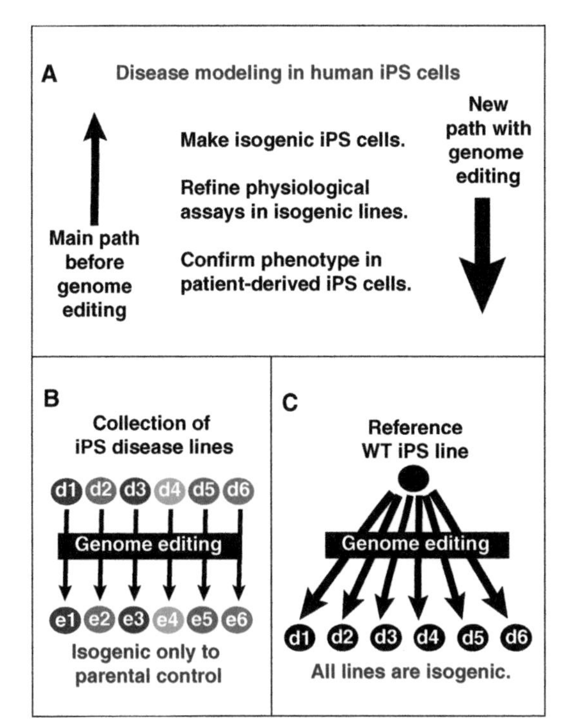

FIGURE 1. Examples of how genome editing has changed the landscape for phenotyping human iPS cells. (*A*) Before genome-editing methods became robust, scientists made iPS-cell disease lines and attempted to develop in vitro assays with imperfect nonisogenic controls (up arrow). Prior to 2009, making isogenic controls was very difficult; it was only done in a few cases. Now that genome-editing methods are robust (larger down arrow), scientists can start with isogenic controls, making assay development much easier. Once the assays are developed, they can be used on the more challenging patient-derived iPS-cell lines. (*B*) Illustration of how collections of patient-derived iPS-cell disease lines derived from patients (d1–d6) can be genome-engineered (e1–e6), providing valuable controls. However, these collections of engineered iPS cells have different genetic backgrounds (indicated by the different colors), so they do not provide an isogenic allelic series. (*C*) Illustration of how using genome engineering on a common reference line results in an allelic series with a common genetic background. Although the number of engineered lines is the same in *B* and *C*, engineering a common reference line (*C*) is much easier, because a single robust iPS line, rather than multiple patient lines (each with potentially different culture conditions), can be chosen. In addition, there is scientific benefit in having an isogenic allelic series. WT, wild type.

than 100,000 lines currently planned for cell banks around the world (Soares et al. 2014). These collections will certainly be of value in the future. However, it is now possible to consider parallel collections of iPS cells that are made from common reference cell lines (Fig. 1C). The advantage of this approach is that each disease line is isogenic with other disease lines and provides an allele series. Furthermore, because multiple gene-editing events can be achieved, gene–gene interactions can be examined to allow for complex traits to be observed on a controlled, isogenic background. This approach is similar to the large collections of gene-targeted mice that have been made on common genetic backgrounds. Although the reference banks of patient-derived iPS cells will have value, the rapid advance of precise genome-editing methods will provide a complementary approach for each investigator who is embarking on a study with iPS cells.

In summary, we introduced a method that will empower scientists to make and detect scarless single-base mutations in many cell types, including human iPS cells. This method provides a wealth of new opportunities to exploit advantages of genome-editing tools and use iPS cells for disease modeling.

REFERENCES

Chen F, Pruett-Miller SM, Huang Y, Gjoka M, Duda K, Taunton J, Collingwood TN, Frodin M, Davis GD. 2011. High-frequency genome editing using ssDNA oligonucleotides with zinc-finger nucleases. *Nat Methods* **8:** 753–755.

da Cunha Santos G, Shepherd FA, Tsao MS. 2011. EGFR mutations and lung cancer. *Annu Rev Pathol* **6:** 49–69.

Ding Q, Lee YK, Schaefer EA, Peters DT, Veres A, Kim K, Kuperwasser N, Motola DL, Meissner TB, Hendriks WT, et al. 2013. A TALEN genome-editing system for generating human stem cell-based disease models. *Cell Stem Cell* **12:** 238–251.

Fu Y, Foden JA, Khayter C, Maeder ML, Reyon D, Joung JK, Sander JD. 2013. High-frequency off-target mutagenesis induced by CRISPR–Cas nucleases in human cells. *Nat Biotechnol* **31:** 822–826.

Gaj T, Gersbach CA, Barbas CF III. 2013. ZFN, TALEN, and CRISPR/Cas-based methods for genome engineering. *Trends Biotechnol* **31:** 397–405.

Giudice A, Trounson A. 2008. Genetic modification of human embryonic stem cells for derivation of target cells. *Cell Stem Cell* **2:** 422–433.

Gupta A, Meng X, Zhu LJ, Lawson ND, Wolfe SA. 2011. Zinc finger protein-dependent and -independent contributions to the in vivo off-target activity of zinc finger nucleases. *Nucleic Acids Res* **39:** 381–392.

Hindson BJ, Ness KD, Masquelier DA, Belgrader P, Heredia NJ, Makarewicz AJ, Bright IJ, Lucero MY, Hiddessen AL, Legler TC, et al. 2011. High-throughput droplet digital PCR system for absolute quantitation of DNA copy number. *Anal Chem* **83:** 8604–8610.

Hsu PD, Scott DA, Weinstein JA, Ran FA, Konermann S, Agarwala V, Li Y, Fine EJ, Wu X, Shalem O, et al. 2013. DNA targeting specificity of RNA-guided Cas9 nucleases. *Nat Biotechnol* **31:** 827–832.

Mali P, Aach J, Stranges PB, Esvelt KM, Moosburner M, Kosuri S, Yang L, Church GM. 2013. CAS9 transcriptional activators for target specificity screening and paired nickases for cooperative genome engineering. *Nat Biotechnol* **31:** 833–838.

Miyaoka Y, Chan AH, Judge LM, Yoo J, Huang M, Nguyen TD, Lizarraga PP, So PL, Conklin BR. 2014. Isolation of single-base genome-edited human iPS cells without antibiotic selection. *Nat Methods* **11:** 291–293.

Miyaoka Y, Chan AH, Conklin BR. 2016. Using digital polymerase chain reaction to detect single-nucleotide substitutions induced by genome editing. *Cold Spring Harb Protoc* doi:10.1101/pdb.prot086801.

Moore JR, Leinwand L, Warshaw DM. 2012. Understanding cardiomyopathy phenotypes based on the functional impact of mutations in the myosin motor. *Circ Res* **111:** 375–385.

Pattanayak V, Ramirez CL, Joung JK, Liu DR. 2011. Revealing off-target cleavage specificities of zinc-finger nucleases by in vitro selection. *Nat Methods* **8:** 765–770.

Ran FA, Hsu PD, Lin CY, Gootenberg JS, Konermann S, Trevino AE, Scott DA, Inoue A, Matoba S, Zhang Y, et al. 2013. Double nicking by RNA-guided CRISPR Cas9 for enhanced genome editing specificity. *Cell* **154:** 1380–1389.

Soares FA, Sheldon M, Rao M, Mummery C, Vallier L. 2014. International coordination of large-scale human induced pluripotent stem cell initiatives: Wellcome Trust and ISSCR workshops white paper. *Stem Cell Reports* **3:** 931–939.

Soldner F, Laganiere J, Cheng AW, Hockemeyer D, Gao Q, Alagappan R, Khurana V, Golbe LI, Myers RH, Lindquist S, et al. 2011. Generation of isogenic pluripotent stem cells differing exclusively at two early onset Parkinson point mutations. *Cell* **146:** 318–331.

Spencer CI, Baba S, Nakamura K, Hua EA, Sears MA, Fu CC, Zhang J, Balijepalli S, Tomoda K, Hayashi Y, et al. 2014. Calcium transients closely reflect prolonged action potentials in iPSC models of inherited cardiac arrhythmia. *Stem Cell Reports* **3:** 269–281.

Takahashi K, Tanabe K, Ohnuki M, Narita M, Ichisaka T, Tomoda K, Yamanaka S. 2007. Induction of pluripotent stem cells from adult human fibroblasts by defined factors. *Cell* **131:** 861–872.

Yu J, Vodyanik MA, Smuga-Otto K, Antosiewicz-Bourget J, Frane JL, Tian S, Nie J, Jonsdottir GA, Ruotti V, Stewart R, et al. 2007. Induced pluripotent stem cell lines derived from human somatic cells. *Science* **318:** 1917–1920.

Cite this introduction as *Cold Spring Harb Protoc*; doi:10.1101/pdb.top090845

Protocol 1

Using Digital Polymerase Chain Reaction to Detect Single-Nucleotide Substitutions Induced by Genome Editing

Yuichiro Miyaoka,[1] Amanda H. Chan,[1] and Bruce R. Conklin[1,2,3]

[1]*Gladstone Institute of Cardiovascular Disease, San Francisco, California 94158;* [2]*Departments of Medicine, and Cellular and Molecular Pharmacology, University of California, San Francisco, California 94143*

This protocol is designed to detect single-nucleotide substitutions generated by genome editing in a highly sensitive and quantitative manner. It uses a combination of allele-specific hydrolysis probes and a new digital polymerase chain reaction (dPCR) technology called droplet digital PCR (ddPCR). ddPCR partitions a reaction into more than 10,000 nanoliter-scale water-in-oil droplets. As a result, each droplet contains only a few copies of the genome so that ddPCR is able to detect rare genome-editing events without missing them.

MATERIALS

It is essential that you consult the appropriate Material Safety Data Sheets and your institution's Environmental Health and Safety Office for proper handling of equipment and hazardous materials used in this protocol.

Reagents

ddPCR Buffer Control Kit (Bio-Rad 1863052)
ddPCR Supermix for Probes (no dUTP) (Bio-Rad 1863024)
Droplet Generation Oil for Probes (Bio-Rad 1863005)
Genomic DNA isolated from cells treated with genome-editing tools to induce single-nucleotide substitutions, diluted to 100–150 ng/µL in distilled water or TE buffer (if original concentration is higher)
Positive-control plasmids with the original and changed allelic sequences

Alternatively, allele-specific DNA fragments may be used. These DNA fragments should be contain all the primer and probe binding sequences.

Equipment

DG8 Cartridge Holder (Bio-Rad 1863051)
DG8 Cartridges for Droplet Generator (Bio-Rad 1864008)
DG8 Gaskets for QX200 Droplet Generator (Bio-Rad 1863009)
PCR eight-tube strips (optional; see Step 17)
PCR plates (Eppendorf twin.tec; 96-wells; semiskirted) (Fisher 951020346)
Pierceable Foil Heat Seal (Bio-Rad 1814040)
Pipettes (eight-channel; 20- and 50-µL)
Primer Express Software 3.0 (Life Technologies 4363991)
PX1 PCR Plate Sealer (Bio-Rad 1814000)

[3]Correspondence: bconklin@gladstone.ucsf.edu

Copyright © Cold Spring Harbor Laboratory Press; all rights reserved
Cite this protocol as *Cold Spring Harb Protoc;* doi:10.1101/pdb.prot086801

QX100 or QX200 Droplet Digital PCR system (Bio-Rad 1864001)

The ddPCR system includes a droplet generator and droplet analyzer.

Thermocycler capable of holding 96-well plates

METHOD

Preparation of Hydrolysis Probes and Primers

Allele-specific hydrolysis probes (e.g., TaqMan) and primers (Fig. 1A) are required for this procedure. They may be obtained by following Steps 1–3 or by sending target sequences to a vendor (e.g., Integrated DNA Technologies).

1. Use Primer Express Software 3.0, following the manufacturer's instructions for "TaqMan MGB Allelic Discrimination." Design probes and primers by treating the single-nucleotide substitution as a single-nucleotide polymorphism (SNP) in the software.
2. Select and order a primer pair such that at least one of the two primers binds outside of the donor oligo-DNA sequence (Fig. 1A).

   Typically the amplicon size is <200 bp.
3. Order two allele-specific MGB probes conjugated with fluorescent dyes FAM and VIC/HEX, respectively, from Life Technologies.
4. Prepare two 100-µL hydrolysis probe and primer mixtures, each of which is specific for one allele.

Forward primer (allele-specific; 100 µM)	18 µL
Reverse primer (allele-specific; 100 µM)	18 µL
Probe (FAM or VIC/HEX; 100 µM)	5 µL
Distilled water	59 µL

   Store the mixtures for up to 1 yr at −20°C.

Validation of Probes and Primers for the ddPCR System

This method is adapted from Hindson et al. (2011).

5. Dilute the positive control as appropriate for their size (e.g., to 0.5 pg/µL for a 3000-bp plasmid or to 0.05 pg/µL for a 300-bp DNA fragment) and combine them to make a 1:1 mixture.

   Handle positive-control mixtures carefully. If a reagent is contaminated with this mixture, it will completely disrupt the assay.
6. Assemble a master mix on ice by combining the following reagents (volumes are indicated for one 25-µL reaction).

Distilled water	9 µL
ddPCR Supermix for Probes (no dUTP)	12.5 µL
FAM probe and primer mixture (from Step 4)	1.25 µL
VIC/HEX probe and primer mixture (from Step 4)	1.25 µL
1:1 plasmid mixture (from Step 5)	1 µL

7. Carefully apply 20 µL of the mixture into each of the eight "sample" wells of a DG8 Cartridge for Droplet Generator.

   Bubbles floating on the surface of the sample do not affect droplet generation, but bubbles in the bottom of the well must be removed. Take note of orientation of the cartridge with respect to the sample order to ensure correct loading of the final Eppendorf twin.tec 96-well plate.
8. Apply 70 µL of Droplet Generation Oil for Probes into each of the eight "oil" wells of the DG8 Cartridge for Droplet Generator.

   Do not load the oil before the samples as this will reduce the droplet number.
9. Hook a DG8 Gasket for QX200 Droplet Generator onto the DG8 Cartridge Holder.

Cite this protocol as *Cold Spring Harb Protoc*; doi:10.1101/pdb.prot086801

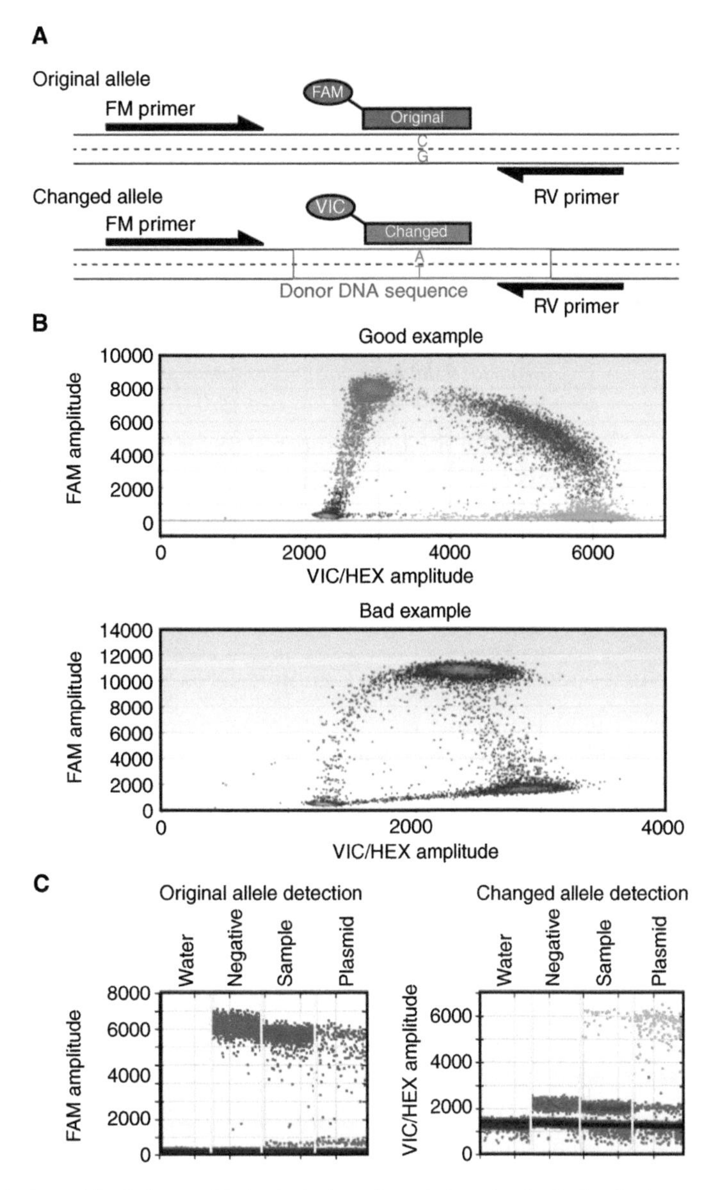

FIGURE 1. (*A*) PCR design with allele-specific hydrolysis probes. A C-to-A substitution is shown as an example. These two alleles are discriminated by different fluorophores; hydrolysis probes specific to the original and changed alleles are conjugated to FAM and VIC (or HEX; not shown), respectively. The primers, however, are identical. One of the two primers hybridizes to a region outside of the donor DNA sequence to amplify the correctly recombined allele. (*B*) Good and bad examples of primer and probe sets to detect single-nucleotide substitutions. In the good example, distinct negative (black), FAM-positive (blue), VIC/HEX-positive (green), and double-positive (brown) populations are clearly seen. However, in the bad example, the FAM-positive and double-positive populations are fused into one population. In such cases, the probes and/or primers must be redesigned. (*C*) Example of ddPCR analysis to detect a single-nucleotide substitution. The original (blue) and changed (green) alleles are detected by the FAM and VIC/HEX signals, respectively. The distilled water control indicates that the system was free from noise, and the negative control indicates that genome-editing tools that targeted an unrelated genomic region did not induce aberrant HDR at the target locus. The positive-plasmid control indicates the amplitudes of the two alleles. In this case, 1.45% of the sample had the changed allele.

10. Put the holder in the droplet generator to generate droplets in eight wells.
11. Transfer droplets into a semiskirted Eppendorf twin.tec 96-well plate using an eight-channel pipette set to 45 µL.

    Do not press the pipette tightly to the bottom of the cartridge or pipette too vigorously as this will shear the droplets. Cover the PCR plate with a foil sheet immediately after transfer to reduce the risk of contamination.

12. Seal the plate with a Pierceable Foil Heat Seal using the PX1 PCR Plate Sealer set to 180°C.
13. Perform thermal cycling using the following program.

1 cycle	95°C	10 min
40 cycles	94°C	30 sec
	50°C–60°C gradient	1 min
1 cycle	98°C	10 min
1 cycle	12°C	hold

14. Analyze the droplets using the ddPCR analyzer.
    i. Go to "Analyze" and then "2-D Amplitude."

    In a successful assay, distinct negative, FAM-positive, VIC/HEX-positive, and double-positive populations should be seen in the 2D plot (Fig. 1B). See Troubleshooting.

    ii. Determine the best temperature for annealing and extension by identifying the highest temperature at which the best separation of the four populations is achieved.
15. (Optional) Make a dilution series of the changed allele to determine the sensitivity of the ddPCR assay for the target.

    A typical dilution series is 0.01%, 0.1%, 1%, 10%, 50%, and 100% of the changed allele over the original allele. Repeat Steps 5–14 with a dilution series to determine the limit of detection.

Detection of Single-Nucleotide Substitutions in Genomic-DNA Samples

16. Assemble a master mix on ice by combining the following reagents. Prepare a total volume sufficient for every sample as well as a distilled water–only negative control and a 1:1 positive-control mixture.

Distilled water	9 µL
ddPCR Supermix for Probes (no dUTP)	12.5 µL
FAM probe and primer mixture (from Step 4)	1.25 µL
VIC/HEX probe and primer mixture (from Step 4)	1.25 µL

In Step 18, 1 µL of 100–150 ng/µL genomic DNA will be added to bring the final volume of each reaction to 25 µL. If using a different volume of genomic DNA in Step 18, adjust the volume of distilled water used here accordingly.

17. Aliquot 24 µL of master mix from Step 16 into PCR eight-tube strips or any 96-well plate.
18. Add 1 µL (100–150 ng) of genomic DNA per sample. Also, add controls to tubes or wells designated for the controls.
19. If the total number of samples to be analyzed is not a multiple of eight, fill the remaining empty tubes of the PCR strip with 25 µL of ddPCR Buffer Control Kit diluted to 1× with distilled water.
20. Briefly centrifuge the PCR tubes or plates.
21. Gently pipette up and down to mix the reactions using a 20-µL eight-channel pipette, and carefully apply the mixtures into "sample" wells.
22. Repeat Steps 8–11 until droplets are generated for all the samples.
23. Seal the plate with a Pierceable Foil Heat Seal using the PX1 PCR Plate Sealer set to 180°C.
24. Perform thermal cycling with the optimal temperature identified in Step 14.ii.

1 cycle	95°C	10 min
40 cycles	94°C	30 sec
	temperature from Step 14.ii	1 min
1 cycle	98°C	10 min
1 cycle	12°C	hold

25. Analyze the droplets using the ddPCR analyzer (Fig. 1C).
    i. Go to "Analyze" and then "2D Amplitude."

Cite this protocol as *Cold Spring Harb Protoc*; doi:10.1101/pdb.prot086801

ii. Gate negative, FAM-positive, VIC/HEX-positive, and double-positive populations by comparing samples with controls.
iii. Go to "concentration" to obtain the allelic frequencies.
iv. Export data as a CSV file by "Export CSV."

TROUBLESHOOTING

Problem (Step 14.i): Four distinct populations cannot be detected (e.g., see Fig. 1B).
Solution: The 50°C–60°C gradient may not yield optimal separation of the four populations; performing the probe/primer validation again using an adjusted temperature gradient may solve the problem. Alternatively, the probes and/or primers may have to be redesigned.

RELATED INFORMATION

For background information on this protocol, see Miyaoka et al. (2014) and Introduction: Detecting Single-Nucleotide Substitutions Induced by Genome Editing (Miyaoka et al. 2016).

REFERENCES

Hindson BJ, Ness KD, Masquelier DA, Belgrader P, Heredia NJ, Makarewicz AJ, Bright IJ, Lucero MY, Hiddessen AL, Legler TC, et al. 2011. High-throughput droplet digital PCR system for absolute quantitation of DNA copy number. *Anal Chem* **83:** 8604–8610.

Miyaoka Y, Chan AH, Judge LM, Yoo J, Huang M, Nguyen TD, Lizarraga PP, So PL, Conklin BR. 2014. Isolation of single-base genome-edited human iPS cells without antibiotic selection. *Nat Methods* **11:** 291–293.

Miyaoka Y, Chan AH, Conklin BR. 2016. Detecting single-nucleotide substitutions induced by genome editing. *Cold Spring Harb Protoc* doi: 10.1101/pdb.top090845.

CHAPTER 7

CRISPR–Cas9 Genome Engineering in *Saccharomyces cerevisiae* Cells

Owen W. Ryan,[1] Snigdha Poddar,[1] and Jamie H.D. Cate[1,2,3,4]

[1]*Energy Biosciences Institute, University of California, Berkeley, California 94720;* [2]*Department of Molecular and Cell Biology, University of California, Berkeley, California 94720;* [3]*Department of Chemistry, University of California, Berkeley, California 94720*

This protocol describes a method for CRISPR–Cas9-mediated genome editing that results in scarless and marker-free integrations of DNA into *Saccharomyces cerevisiae* genomes. DNA integration results from cotransforming (1) a single plasmid (pCAS) that coexpresses the Cas9 endonuclease and a uniquely engineered single guide RNA (sgRNA) expression cassette and (2) a linear DNA molecule that is used to repair the chromosomal DNA damage by homology-directed repair. For target specificity, the pCAS plasmid requires only a single cloning modification: replacing the 20-bp guide RNA sequence within the sgRNA cassette. This CRISPR–Cas9 protocol includes methods for (1) cloning the unique target sequence into pCAS, (2) assembly of the double-stranded DNA repair oligonucleotides, and (3) cotransformation of pCAS and linear repair DNA into yeast cells. The protocol is technically facile and requires no special equipment. It can be used in any *S. cerevisiae* strain, including industrial polyploid isolates. Therefore, this CRISPR–Cas9-based DNA integration protocol is achievable by virtually any yeast genetics and molecular biology laboratory.

MATERIALS

It is essential that you consult the appropriate Material Safety Data Sheets and your institution's Environmental Health and Safety Office for proper handling of equipment and hazardous material used in this protocol.

RECIPES: Please see the end of this protocol for recipes indicated by <R>. Additional recipes can be found online at http://cshprotocols.cshlp.org/site/recipes.

Reagents

Agarose gel, 2% in TAE <R>

We use SYBR Safe DNA Gel Stain (1:10,000 dilution) directly in the gels.

DNA ladder, GeneRuler 1 kb Plus (Life Technologies SM1331)
DNA oligonucleotides (see Steps 1–2, 10, and 24–55)
DNA, single stranded from salmon testes (ssDNA) (Sigma-Aldrich D7656)
dNTPs (10 mM total; 2.5 mM each)
DpnI restriction enzyme and supplied 10× digestion buffer (New England BioLabs R0176S)
Escherichia coli chemically competent cells (Life Technologies C4040-10)
ExoSAP-IT (Affymetrix 78250)

[4]Correspondence: jcate@lbl.gov

Copyright © Cold Spring Harbor Laboratory Press; all rights reserved
Cite this protocol as *Cold Spring Harb Protoc*; doi:10.1101/pdb.prot086827

GeneMorph II error-prone PCR kit (Agilent 200550)
Glycerol (50%)
LATE solution <R>
LB (Luria–Bertani) liquid medium <R>
LB+ kanamycin (100 mg/L)
LB+ kanamycin (100 mg/L), agar (2%) plates
Loading dye, 6× (Thermo Scientific R0611)
Miniprep kit (QIAGEN)
pCAS plasmid (Addgene plasmid 60847)
Phusion High-Fidelity DNA Polymerase kit (New England Biolabs M0530L)
PLATE transformation solution, freshly prepared <R>
Sterile water

Use for all steps requiring water as a reagent.

Yeast cells of interest
YPD <R>
YPD, agar (2%) plates
YPD+ G418 (200 mg/L), agar (2%) plates

Equipment

Falcon tubes, 50 mL
Freezer, −80°C
Gel imager
Heat block or water bath, 42°C
Incubators, 37°C and 30°C
Microcentrifuge
Microcentrifuge tubes, 1.5 mL
NanoDrop 1000 spectrophotometer
PCR tubes
Shake flasks, baffled
Spectrophotometer
Tabletop centrifuge (Eppendorf 5810R)
Thermocycler

METHOD

Cloning the Guide Sequence into the pCAS sgRNA Cassette

See Figure 1 for a diagram of the sgRNA cassette in the pCAS plasmid (Ryan et al. 2014).

1. Design a 60-mer oligonucleotide for cloning the unique guide sequence with the following sequence:

   5′-CGGGTGGCGAATGGGACTTT-[20-bp target sequence]-GTTTTAGAGCTAGAAATAGC-3′

   The unique 20-bp target sequence must reside immediately 5′ of the PAM sequence (NGG) within the yeast genome. Avoid stretches of four or more thymine (T) nucleotides within the 20-bp target sequence, which

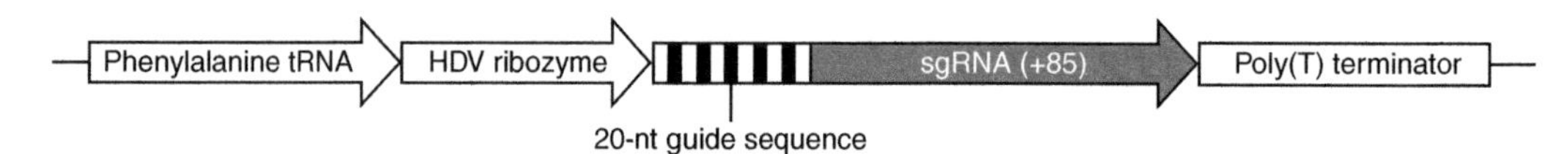

FIGURE 1. Architecture of the sgRNA cassette in the pCAS plasmid. The 20-bp DNA target sequence (vertical bars) immediately 5′ of an NGG protospacer adjacent motif (PAM) sequence is cloned as a guide sequence into a universal sgRNA cassette.

Cite this protocol as *Cold Spring Harb Protoc*; doi:10.1101/pdb.prot086827

can serve as an RNA polymerase III terminator (Braglia et al. 2005). There are other suitable tRNA promoters to drive the sgRNA in nonlaboratory strains of S. cerevisiae (Ryan et al. 2014).

2. Order two 60-mer oligonucleotides, one with the sgRNA guide coding sequence (Step 1) and one with the reverse complement of the coding sequence.
3. Dilute each of the two 60-mers to 100 µM in water.
4. Assemble the reaction in a polymerase chain reaction (PCR) tube:

Phusion HF buffer (5×)	5 µL
dNTP (10 mM total; 2.5 mM each)	0.5 µL
Coding sgRNA guide sequence (60-mer; 100 µM)	0.1 µL
Reverse complement of sgRNA (60-mer; 100 µM)	0.1 µL
pCAS plasmid	40 ng
Phusion DNA Polymerase	1 µL
H_2O	to 25 µL

Phusion HF buffer and DNA polymerase are provided in the Phusion High-Fidelity DNA Polymerase kit.

5. Perform thermocycling with the following profile.

1 cycle	98°C	1 min
30 cycles	98°C	30 sec
	58°C	1 min
	72°C	10 min
1 cycle	72°C	10 min
	4°C	Hold

6. Add 1 µL of DpnI, 3 µL of the provided 10× digestion buffer, and 1 µL of water to the cloning reaction and incubate for 6 h at 37°C (or overnight).
7. Transform the DpnI-treated reaction into *E. coli* competent cells:
   i. Add 15 µL of the cloning reaction into 50 µL of ice-thawed competent bacterial cells.
   ii. Incubate on ice for 30 min.
   iii. Heat shock in a heat block or water bath for 1 min at 42°C.
   iv. Recover in 200 µL of LB liquid medium for 1 h at 37°C.
   v. Spread the contents of the reaction onto LB+ kanamycin plates.
   vi. Incubate overnight at 37°C.
8. Pick five to 10 colonies from the LB+ kanamycin plate and shake-incubate overnight in 3 mL of LB+ kanamycin at 37°C.
9. Isolate the plasmids using a QIAGEN Miniprep Kit and quantify DNA using the NanoDrop spectrophotometer (1 A_{260} OD = 50 ng/µL).
10. Sequence the sgRNA construct by Sanger sequencing with the sequencing primer 5′-CGGAA TAGGAACTTCAAAGCG-3′.

Making Competent Yeast Cells

11. Inoculate yeast cells from a single colony in 3 mL of YPD and incubate overnight at 30°C with constant shaking at 220 rpm.
12. Dilute the overnight yeast culture to an OD_{600} of 0.2 in 100 mL of YPD in a baffled shake flask.
13. Grow culture at 30°C shaking at 220 rpm until $OD_{600} = 1.0$.
14. Divide culture evenly into 50-mL Falcon tubes.

    Perform Steps 15–23 for each Falcon tube.
15. Centrifuge at 3000 rpm (1400*g*) for 2 min.

16. Remove supernatant.
17. Resuspend in 0.5 mL of LATE solution.
18. Transfer to a 1.5-mL microcentrifuge tube.
19. Centrifuge at 10,000 rpm (8600*g*) for 1 min.
20. Remove supernatant.
21. Resuspend pellet in 300 µL LATE solution.
22. Add 300 µL 50% glycerol.
23. Pipette mix and store at −80°C.

Generating Double-Stranded DNA Repair

See Figure 2 for schematics of the dsDNA repair strategies described below. Follow Steps 24–29 for assembling a DNA barcode (Fig. 2A), Steps 30–35 for generating a mutation within 20 nucleotides 5′ of the PAM sequence (Fig. 2B), Steps 36–41 for generating a point mutation within 60 nucleotides 5′ of the PAM (Fig. 2C), Steps 42–46 for generating synthetic gene constructs for in vivo DNA assembly (Fig. 2D), and Steps 47–55 for generating DNA libraries by error-prone PCR (Fig. 2E).

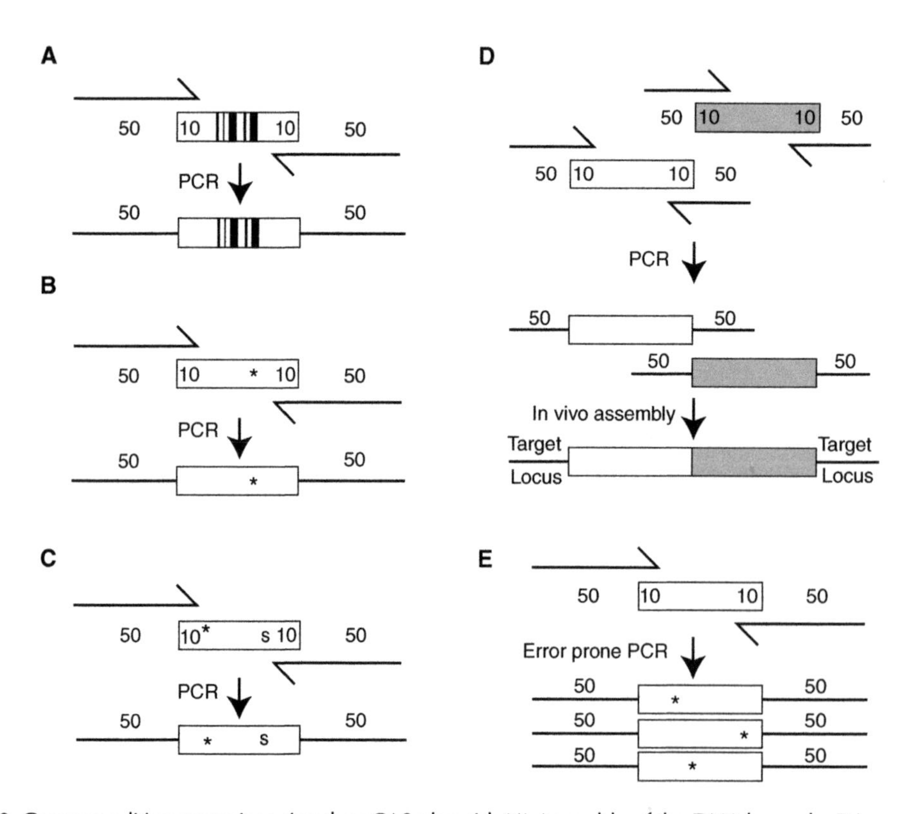

FIGURE 2. Genome-editing strategies using the pCAS plasmid. (*A*) Assembly of the DNA barcode. Primers consist of 10 bp of barcode priming sequence and 50 bp homologous to the genome target locus. (*B*) Insertion alleles with a mutation within 20 bp of a PAM sequence. The 60-mer insertion allele oligo is identical to the target locus but contains the desired mutation (*). Primers contain 50 bp of homology with the genome target locus. (*C*) Mutant alleles with a mutation further than 20 bp from a PAM. The 60-mer insertion allele oligo is identical to the target locus but contains the desired mutation (*) and a synonymous substitution within 20 bp of the PAM (s). Primers contain 50 bp of homology with the target locus. (*D*) In vivo DNA assembly. Flanking primers contain 50 bp of homology with the genome target locus, and internal primers contain 50 bp of homology with the adjacent insertion fragment. (*E*) Mutant allele libraries. Primers contain 50 bp of homology with the target locus. Error-prone PCR is used to generate a library of mutant alleles (*).

Cite this protocol as *Cold Spring Harb Protoc*; doi:10.1101/pdb.prot086827

Assembling a DNA Barcode (Fig. 2A)

24. Order three 60-mer oligonucleotides:
    Barcode (with 5′ TAA stop codon): 5′-TAAGATGTCC-[40-bp barcode]-GCAGCGTACG-3′
    Forward primer: 5′-[50 bp homology]-TAAGATGTCC-3′
    Reverse primer: 5′-[50 bp homology]-CGTACGCTGC-3′
25. Dilute all oligos to 100-µM stock concentrations.
26. Dilute the barcode oligo to 100 nM in water.
27. Assemble the PCR mixture:

H_2O	38.45 µL
Phusion HF buffer (5×)	10 µL
dNTP (10 mM total; 2.5 mM each)	1 µL
Forward primer (100 µM)	0.1 µL
Reverse primer (100 µM)	0.1 µL
Phusion DNA Polymerase	0.25 µL
Barcode oligo	0.1 µL

Phusion HF buffer and DNA polymerase are provided in the Phusion High-Fidelity DNA Polymerase kit.

28. Perform thermocycling with the following profile.

1 cycle	98°C	1 min
30 cycles	98°C	30 sec
	50°C	1 min
	72°C	1 min
1 cycle	72°C	7 min
	4°C	hold

29. Run 5 µL of each PCR on a 2% agarose TAE gel with 1 µL of 6× loading dye. Include the DNA ladder as a size standard.
    The PCR product can be transformed into yeast cells (no purification required); see Steps 56–70.

Generating a Mutation within 20 Nucleotides 5′ of the PAM Sequence (Fig. 2B)

30. Design an sgRNA sequence for the DNA that is to be targeted and replaced, using the wild-type allele sequence.
31. Clone the guide sequence into the pCAS plasmid (Steps 1–10).
32. Order a 60-mer oligo of the DNA sequence to be integrated (mutant allele sequence).
33. Design primers to amplify the insertion allele containing 50 bp of homology with the genome target.
34. PCR-amplify the mutant allele repair DNA oligonucleotide using the same thermocycler program as for barcode assembly (Step 28).
35. Run 5 µL of each PCR on a 2% agarose TAE gel with 1 µL of 6× loading dye. Include the DNA ladder as a size standard.
    The PCR product can be transformed into yeast cells (no purification required); see Steps 56–70.

Generating a Point Mutation within 60 Nucleotides 5′ of the PAM (Fig. 2C)

36. Design an sgRNA sequence for the DNA that is to be targeted using the wild-type allele sequence.
37. Clone the guide sequence into the pCAS plasmid (Steps 1–10).
38. Order a 60-mer oligonucleotide of the DNA sequence to be integrated. Ensure that it contains two mutations: (1) the mutation to be integrated and (2) a synonymous codon substitution within 20 nucleotides 5′ of the PAM.
    Mutations can be generated >60 bp 5′ of the PAM, but longer oligos need to be purchased.

39. Design primers to amplify the insertion allele containing 50 bp of homology with the genome target.
40. PCR-amplify the mutant allele repair DNA oligonucleotide using the same thermocycler program as for barcode assembly (Step 28).
41. Run 5 µL of each PCR on a 2% agarose TAE gel with 1 µL of 6× loading dye. Include the DNA ladder as a size standard.

    The PCR product can be transformed into yeast cells (no purification required); see Steps 56–70.

Generating Synthetic Gene Constructs for In Vivo DNA Assembly (Fig. 2D)

42. Design an sgRNA sequence for the DNA that is to be targeted for insertion or replacement, using the wild-type allele sequence.
43. Clone the guide sequence into the pCAS plasmid (Steps 1–10).
44. Design amplification primers for each DNA fragment used to encode the gene of interest.

    For overlapping DNA fragments, flanking primers contain 50 bp of homology with the target site while internal primers contain 50 bp of homology with the intended adjacent fragment.

45. PCR-amplify DNA fragments from plasmid or gDNA using standard high-fidelity PCR techniques.
46. Run 5 µL of each PCR on a 2% agarose TAE gel with 1 µL of 6× loading dye. Include the DNA ladder as a size standard.

    The PCR product can be transformed into yeast cells (no purification required); see Steps 56–70.

Generating DNA Libraries by Error-Prone PCR (Fig. 2E)

47. Design primers to amplify the DNA with 50 bp of homology with the target genome.
48. Amplify the DNA that is to serve as the template for mutant library preparation, using standard high-fidelity PCR techniques.
49. Add 1 µL of ExoSAP-IT to the PCR and incubate for 30 min.
50. Quantify DNA using a NanoDrop 1000 spectrophotometer (1 A_{260} OD = 50 ng/µL).
51. Run 5 µL of each PCR on a 2% agarose TAE gel with 1 µL of 6× loading dye. Include the DNA ladder as a size standard.
52. Use the appropriate amount of PCR product as error-prone PCR template to achieve the desired mutation rate.
53. Assemble the PCR mixture.

H_2O	to 50 µL
GeneMorph II buffer (10×)	10 µL
dNTP (40 mM from GeneMorph II kit)	1 µL
Forward primer (100 µM)	0.2 µL
Reverse primer (100 µM)	0.2 µL
GeneMorph II enzyme	1.0 µL
PCR DNA template	*x* µL

GeneMorph II buffer and enzyme are provided in the GeneMorph II error-prone PCR kit.

54. Perform thermocycling with the following profile.

1 cycle	95°C	1 min
30 cycles	95°C	30 sec
	50°C	30 sec
	72°C	1 min (1 kb per minute)
1 cycle	72°C	10 min
	4°C	Hold

Cite this protocol as *Cold Spring Harb Protoc*; doi:10.1101/pdb.prot086827

55. Run 5 µL on a 2% agarose TAE gel with 1 µL of 6× loading dye. Include the DNA ladder as a size standard.

    The PCR product can be transformed into yeast cells (no purification required); see Steps 56–70.

    Libraries can be combined in vivo with overlapping PCR fragments but transformation efficiency (library size) decreases with increased number of DNA fragments.

Cotransformation of pCAS and Linear DNA into Yeast Cells

56. Place salmon testes ssDNA tube in boiling water for 5 min and immediately place it on ice for 5 min.
57. Thaw competent cells (from Step 23) for 10 min at room temperature.
58. Combine 1.0 µg of pCAS plasmid, 5.0 µg of linear repair DNA, 10 µL of ssDNA, 90 µL of yeast competent cells, and 900 µL of PLATE solution in a 1.5-mL microcentrifuge tube and pipette to mix.
59. Incubate for 30 min at 30°C.
60. Heat shock for 15 min at 42°C.
61. Centrifuge at 10,000 rpm (8600*g*) for 1 min.
62. Use a pipette to aspirate supernatant.
63. Resuspend cells in 250 µL of YPD.
64. Recover for 2.0 h at 30°C in the microcentrifuge tube (no shaking required).
65. Plate entire contents of the microcentrifuge tube onto YPD + G418 (200 mg/L).
66. Incubate for 48 h at 37°C.
67. Pick single colonies from the YPD + G418 plate and grow overnight in YPD at 30°C and 220 rpm.
68. Streak culture for single colonies on YPD plates (no G418).
69. Streak cells from YPD plate onto YPD + G418 plates to confirm plasmid loss (G418 sensitivity).
70. Confirm correct integration by diagnostic PCR and sequencing.

DISCUSSION

CRISPR–Cas9 has been broadly showed as a facile method for marker-free and multiplex genome engineering of many organisms. Here, we have presented our method for CRISPR–Cas9 genome editing in *S. cerevisiae*, the principle eukaryotic microorganism in synthetic biology and industrial biotechnology. We have described how to clone an sgRNA into a base Cas9 vector and use this recombinant plasmid along with a synthetic, double-stranded linear DNA molecule for genome editing in yeast. The linear DNA molecule is integrated by endogenous DNA repair mechanisms and enables genomic changes such as single-nucleotide polymorphisms, insertions/deletions (indels), knockout mutants, and knock-in mutants. This system functions with high efficiency and fidelity in *S. cerevisiae*, resulting in precise genome editing while not affecting the remainder of the genome (Ryan et al. 2014; Ryan and Cate 2014).

The low cost, technical simplicity, and high efficiency of CRISPR–Cas9 genome-editing methods make them powerful tools for large-scale mutational analyses and synthetic biology experiments, such as elaborate one-step engineering of multigene biosynthetic pathways. Further, random DNA libraries can be cloned directly into yeast genomes, enabling high-throughput in situ directed evolution of proteins or regulatory elements (Ryan et al. 2014). Our modification to the sgRNA format could be combined with other implementations of CRISPR–Cas9 in *S. cerevisiae* (DiCarlo et al. 2013; Bao et al. 2014; Gao and Zhao 2014; Zhang et al. 2014; Horwitz et al. 2015; Jakočiūnas et al. 2015a,b; Lee et al. 2015a,b; Mans et al. 2015; Stovicek et al. 2015; Xu et al. 2015). The ability to rapidly and cost effectively edit yeast genomes should facilitate strain engineering for the synthesis of biofuels, biochemicals, and biopharmaceuticals. We also anticipate that, with minor modifications, this CRISPR–Cas9 system will be portable to other industrially relevant eukaryotic organisms.

RECIPES

LATE Solution

40 mL H_2O
5 mL lithium acetate (1 M)
5 mL TE buffer (10×; 100 mM Tris-Cl, 10 mM EDTA [pH 8.0])

Store for up to 6 mo at room temperature.

LB (Luria-Bertani) Liquid Medium

Reagent	Amount to add
H_2O	950 mL
Tryptone	10 g
NaCl	10 g
Yeast extract	5 g

Combine the reagents and shake until the solutes have dissolved. Adjust the pH to 7.0 with 5 N NaOH (~0.2 mL). Adjust the final volume of the solution to 1 L with H_2O. Sterilize by autoclaving for 20 min at 15 psi (1.05 kg/cm^2) on liquid cycle.

PLATE Transformation Solution

40 mL polyethyleneglycol (PEG) 2000 (50%)
5 mL lithium acetate (1 M)
5 mL TE buffer (10×; 100 mM Tris-Cl, 10 mM EDTA [pH 8.0])

Prepare fresh before transformation.

TAE

Prepare a 50× stock solution in 1 L of H_2O:
242 g of Tris base
57.1 mL of acetic acid (glacial)
100 mL of 0.5 M EDTA (pH 8.0)

The 1× working solution is 40 mM Tris-acetate/1 mM EDTA.

YPD

Peptone, 20 g
Glucose, 20 g
Yeast extract, 10 g
H_2O to 1000 mL
YPD (YEPD medium) is a complex medium for routine growth of yeast.

To prepare plates, add 20 g of Bacto Agar (2%) before autoclaving.

REFERENCES

Bao Z, Xiao H, Liang J, Zhang L, Xiong X, Sun N, Si T, Zhao H. 2014. Homology-integrated CRISPR-Cas (HI-CRISPR) system for one-step multigene disruption in *Saccharomyces cerevisiae*. *ACS Synth Biol* **4:** 585–594.

Braglia P, Percudani R, Dieci G. 2005. Sequence context effects on oligo(dT) termination signal recognition by *Saccharomyces cerevisiae* RNA polymerase III. *J Biol Chem* **280:** 19551–19562.

Cite this protocol as *Cold Spring Harb Protoc*; doi:10.1101/pdb.prot086827

DiCarlo JE, Norville JE, Mali P, Rios X, Aach J, Church GM. 2013. Genome engineering in *Saccharomyces cerevisiae* using CRISPR–Cas systems. *Nucleic Acids Res* **41:** 4336–4343.

Gao Y, Zhao Y. 2014. Self-processing of ribozyme-flanked RNAs into guide RNAs in vitro and in vivo for CRISPR-mediated genome editing. *J Integr Plant Biol* **56:** 343–349.

Horwitz AA, Walter JM, Schubert MG, Kung SH, Hawkins K, Platt DM, Hernday AD, Mahatdejkul-Meadows T, Szeto W, Chandran SS, et al. 2015. Efficient multiplexed integration of synergistic alleles and metabolic pathways in yeasts via CRISPR–Cas. *Cell Systems* **1:** 88–96.

Jakočiūnas T, Bonde I, Herrgård M, Harrison SJ, Kristensen M, Pedersen LE, Jensen MK, Keasling JD. 2015a. Multiplex metabolic pathway engineering using CRISPR–Cas9 in *Saccharomyces cerevisiae. Metab Eng* **28:** 213–222.

Jakočiūnas T, Rajkumar AS, Zhang J, Arsovska D, Rodriguez A, Jendresen CB, Skjødt ML, Nielsen AT, Borodina I, Jensen MK, et al. 2015b. CasEMBLR: Cas9-facilitated multiloci genomic integration of in vivo assembled DNA parts in *Saccharomyces cerevisiae. ACS Synth Biol* **4:** 1226–1234.

Lee ME, DeLoache WC, Cervantes B, Dueber JE. 2015a. A highly characterized yeast toolkit for modular, multipart assembly. *ACS Synth Biol* **4:** 975–986.

Lee NCO, Larionov V, Kouprina N. 2015b. Highly efficient CRISPR/Cas9-mediated TAR cloning of genes and chromosomal loci from complex genomes in yeast. *Nucleic Acids Res* **43:** e55.

Mans R, van Rossum HM, Wijsman M, Backx A, Kuijpers NGA, van den Broek M, Daran-Lapujade P, Pronk JT, van Maris AJA, Daran J-MG. 2015. CRISPR/Cas9: A molecular Swiss army knife for simultaneous introduction of multiple genetic modifications in *Saccharomyces cerevisiae. FEMS Yeast Res* **15:** fov004.

Ryan OW, Cate JHD. 2014. Multiplex engineering of industrial yeast genomes using CRISPRm. *Meth Enzymol* **546:** 473–489.

Ryan OW, Skerker JM, Maurer MJ, Li X, Tsai JC, Poddar S, Lee ME, DeLoache W, Dueber JE, Arkin AP, et al. 2014. Selection of chromosomal DNA libraries using a multiplex CRISPR system. *Elife* **3:** e03703.

Stovicek V, Borodina I, Forster J. 2015. CRISPR–Cas system enables fast and simple genome editing of industrial *Saccharomyces cerevisiae* strains. *Metab Eng Commun* **2:** 13–22.

Xu K, Ren C, Liu Z, Zhang T, Zhang T, Li D, Wang L, Yan Q, Guo L, Shen J, et al. 2015. Efficient genome engineering in eukaryotes using Cas9 from *Streptococcus thermophilus. Cell Mol Life Sci* **72:** 383–399.

Zhang G-C, Kong II, Kim H, Liu J-J, Cate JHD, Jin Y-S. 2014. Construction of a quadruple auxotrophic mutant of an industrial polyploid *Saccharomyces cerevisiae* strain by using RNA-guided Cas9 nuclease. *Appl Environ Microbiol* **80:** 7694–7701.

CHAPTER 8

Cas9-Mediated Genome Engineering in *Drosophila melanogaster*

Benjamin E. Housden[1,3] and Norbert Perrimon[1,2,3]

[1]*Department of Genetics, Harvard Medical School, Boston, Massachusetts 02115;* [2]*Howard Hughes Medical Institute, Harvard Medical School, Boston, Massachusetts 02115*

The recent development of the CRISPR–Cas9 system for genome engineering has revolutionized our ability to modify the endogenous DNA sequence of many organisms, including *Drosophila.* This system allows alteration of DNA sequences in situ with single base-pair precision and is now being used for a wide variety of applications. To use the CRISPR system effectively, various design parameters must be considered, including single guide RNA target site selection and identification of successful editing events. Here, we review recent advances in CRISPR methodology in *Drosophila* and introduce protocols for some of the more difficult aspects of CRISPR implementation: designing and generating CRISPR reagents and detecting indel mutations by high-resolution melt analysis.

INTRODUCTION

The process of genome engineering relies on the generation of a double-strand break (DSB) at a specific, user-defined locus in the genome. Following this event, endogenous cellular mechanisms repair the DNA damage primarily through the nonhomologous end joining (NHEJ) or homologous recombination (HR) pathways (Jeggo 1998; van Gent et al. 2001). NHEJ is an error-prone repair mechanism, often resulting in the generation of short insertion or deletion (indel) mutations at the DSB site. This can be exploited for the generation of gene knockouts by inducing frameshift mutations in coding sequence. Alternatively, by providing a donor construct with homology to the target locus, which can serve as a template for repair by the HR machinery, the repair process can be used to produce more precise alterations to the genomic sequence. For example, single-nucleotide changes, insertions, deletions, or substitutions are possible using HR.

Several existing technologies have been developed to generate DSBs at specific genomic loci (e.g., TALENs [transcription-activator-like effector nucleases] and zinc-finger nucleases) (Bibikova et al. 2003; Joung and Sander 2013; Carroll 2014); however, CRISPR has been widely adopted because of the simplicity of its use and its high efficiency of DSB generation. Since the first demonstrations of genome engineering with CRISPR (Jinek et al. 2012, 2013; Cong et al. 2013; Mali et al. 2013b), the technology has been used in a wide range of organisms, including *Drosophila* (Bassett et al. 2013; Gratz et al. 2013; Kondo and Ueda 2013; Ren et al. 2013; Yu et al. 2013), for a rapidly expanding repertoire of applications.

CRISPR requires two components. The first is a nonspecific DNA nuclease protein called Cas9. The second is an RNA molecule capable of binding to Cas9, which provides customizable specificity to a DNA sequence (Jinek et al. 2012). Although the endogenous CRISPR system in *Streptococcus pyogenes* uses two RNA molecules, these have been combined into a single guide RNA (sgRNA) to

[3]Correspondence: bhousden@genetics.med.harvard.edu; perrimon@receptor.med.harvard.edu

Copyright © Cold Spring Harbor Laboratory Press; all rights reserved

Cite this introduction as *Cold Spring Harb Protoc;* doi:10.1101/pdb.top086843

further simplify the use of the system in other organisms (Jinek et al. 2012). A 20-bp region of the sgRNA provides specificity to the complementary DNA sequence. The target site can be determined simply by recoding this region. Therefore, CRISPR can be directed to different genomic loci just by expressing different sgRNAs. This is in contrast to previous genome engineering technologies based on TALENs and zinc-finger nucleases, which require production of new proteins for each target site. In addition, the only targeting requirement in the CRISPR system is the presence of a PAM (protospacer-adjacent motif) sequence (NGG) 3′ of the 20-bp target site (Jiang et al. 2013); thus, possible target sequences are common. Because CRISPR enables rapid and efficient genome modification with minimal effort required to reprogram target site specificity, it is a highly flexible and robust system for genome engineering.

CRISPR IN *Drosophila*

The use of CRISPR for any application in *Drosophila* requires several steps. First, sgRNA design is critical to achieving the desired editing event at high frequency and with high specificity. Second, CRISPR reagents must be delivered efficiently. Finally, an appropriate method must be available to detect the desired editing event. Here we summarize important factors for the experimental design.

Design of sgRNAs

CRISPR has been shown to function with high efficiency both in cultured *Drosophila* cells and in vivo. Detailed protocols for many aspects of its use have been described elsewhere (Bassett et al. 2013, 2014; Gratz et al. 2013; Kondo and Ueda 2013; Ren et al. 2013; Yu et al. 2013; Bassett and Liu 2014; Beumer and Carroll 2014; Bottcher et al. 2014; Housden et al. 2014). However, despite the widespread adoption of CRISPR methodologies in *Drosophila*, there remain several unanswered questions regarding an optimal experimental design. For example, off-target (OT) mutations have been shown to occur often in mammalian systems, both at predicted OT sites and at other loci not predicted by any current sgRNA design algorithm (Carroll 2013; Fu et al. 2013; Hsu et al. 2013; Mali et al. 2013a; Pattanayak et al. 2013; Tsai et al. 2015). In contrast, OT mutations have yet to be reported in *Drosophila*, perhaps because of reduced genome complexity. In addition, although mutation efficiency varies widely between different sgRNAs, investigations into the difference in efficiencies have produced inconsistent results (Doench et al. 2014; Ren et al. 2014; Wang et al. 2014). It is therefore advisable to design sgRNAs carefully using the available online tools (see Protocol 3: Design and Generation of *Drosophila* Single Guide RNA Expression Constructs [Housden et al. 2016]) and, where possible, perform experiments using multiple sgRNAs in parallel.

Delivery of CRISPR Reagents

The method used to deliver CRISPR reagents into flies can greatly affect the efficiency of genome editing. Delivery methods have been discussed in detail elsewhere (Ren et al. 2013; Housden et al. 2014), so we will provide only a brief summary here. One option is to generate Cas9 mRNA or protein in vitro and inject it together with synthesized sgRNA directly into fly embryos (Gratz et al. 2013; Lee et al. 2014b). This approach can be laborious because of the need to generate high-quality protein or mRNA and generally results in low efficiency. However, one advantage is that the reagents are short-lived and so the chance of generating OT mutations or somatic mutations is reduced, resulting in lower toxicity. An alternative approach is to inject an sgRNA-expressing plasmid into flies that express Cas9 in the germline, which results in higher efficiency compared with direct mRNA injection (Ren et al. 2013). Finally, both sgRNA and Cas9 can be expressed from transgenes. By crossing flies that each express one of these components, very high modification rates can be achieved (Kondo and Ueda 2013). The disadvantages of this approach are the increased time required to generate the transgenic lines and the challenge of delivering a donor for homologous recombination.

 Cite this introduction as *Cold Spring Harb Protoc*; doi:10.1101/pdb.top086843

Delivery of CRISPR reagents into cultured cells can be problematic, as many *Drosophila* cell lines suffer from low transfection efficiency. One option is to encode both sgRNA and Cas9 on a single plasmid (see Protocol 3: Design and Generation of *Drosophila* Single Guide RNA Expression Constructs [Housden et al. 2016]), and transfect the cell type of interest to generate indel mutations. This method works with relatively high efficiency, although when mutations cause cell viability effects, the population can rapidly revert to wild type. Including an antibiotic selection cassette in the CRISPR expression plasmid allows enrichment for transfected cells (Bassett et al. 2014), thereby increasing the proportion of successfully modified cells. However, because of the accumulation of nonframeshift mutations, these populations can also revert to wild type. One solution to this issue is to generate clonal mutant populations by selecting homozygous mutant cells from the mixed population and growing new cultures from individual cells, yet this approach is also problematic because of the low survival rate of single cells in culture.

To stabilize a genomic change in cells, a donor construct or oligo can be cotransfected with the CRISPR expression plasmid to insert a selection cassette into the genome and simultaneously disrupt the gene of interest (see Protocol 1: Design and Generation of Donor Constructs for Genome Engineering in *Drosophila* [Housden and Perrimon 2016a]), leading to more direct selection of the modification event. However, whether donors can integrate nonspecifically, and the frequency of such events, is not yet clear. Although it is possible to include a selection cassette in the HR product, a major limitation is that there is currently no method to ensure all copies of a gene are modified. This is a particular problem in *Drosophila* cell culture because the majority of cell lines are polyploid or aneuploid (Lee et al. 2014a).

Detection of Editing Events

Several methods are available to detect genome-editing events and the approach taken will depend on the type of genomic modification induced. For example, when generating insertions using donor constructs, it is often possible to include a selection marker such as *mini-white* in flies or an antibiotic resistance gene in cells. An alternative approach is to detect insertions using polymerase chain reaction (PCR) assays to amplify fragments present only in the modified flies or cells. In addition, deletions generated using a donor construct can be detected by PCR assays using primers flanking the deletion site.

Detection of small indel mutations can be more challenging because the exact genomic change can be unpredictable and no markers can be included. In this case, molecular assays such as surveyor assays, restriction profiling, or high-resolution melt analysis (HRMA) must be used. Detailed methods for surveyor assays and restriction profiling have been described elsewhere (Bassett et al. 2013; Cong et al. 2013; Housden et al. 2014). We provide a method for HRMA in Protocol 2: Detection of Indel Mutations in *Drosophila* by High-Resolution Melt Analysis (HRMA) (Housden and Perrimon 2016b).

A final consideration for the experimental design is the background in which CRISPR is used. For example, as described above, DSBs can be repaired using NHEJ or HR, with the resulting modification being dependent on which of these pathways is used. To improve the chances of recovering the desired modification, repair pathway choice can be biased. For example, DSBs induced in the background of a *ligase4* mutation have been shown to increase the recovery of HR repair events in vivo and in cells by inhibiting the NHEJ pathway (Beumer et al. 2008; Bozas et al. 2009; Bottcher et al. 2014; Gratz et al. 2014). In addition, small molecules were recently developed to inhibit either HR or NHEJ in pluripotent stem cells (Yu et al. 2015) and thus increase the rate of NHEJ or HR events, respectively. Although these inhibitors have not yet been tested in *Drosophila*, this may provide a useful mechanism to increase efficiency in cell culture.

In summary, the CRISPR system provides a robust and simple method to induce modifications into the *Drosophila* genome. With careful design of the experimental approach and reagents, genome engineering can be achieved with unprecedented efficiency and accuracy.

ACKNOWLEDGMENTS

We thank David Doupé and Stephanie Mohr for helpful discussions during the preparation of this manuscript. Work in the Perrimon Laboratory is supported by grants from the National Institutes of Health and the Howard Hughes Medical Institute.

REFERENCES

Bassett A, Liu JL. 2014. CRISPR/Cas9 mediated genome engineering in *Drosophila*. *Methods* **69:** 128–136.

Bassett AR, Tibbit C, Ponting CP, Liu JL. 2013. Highly efficient targeted mutagenesis of *Drosophila* with the CRISPR/Cas9 system. *Cell Rep* **4:** 220–228.

Bassett AR, Tibbit C, Ponting CP, Liu JL. 2014. Mutagenesis and homologous recombination in *Drosophila* cell lines using CRISPR/Cas9. *Biol Open* **3:** 42–49.

Beumer KJ, Carroll D. 2014. Targeted genome engineering techniques in *Drosophila*. *Methods* **68:** 29–37.

Beumer KJ, Trautman JK, Bozas A, Liu JL, Rutter J, Gall JG, Carroll D. 2008. Efficient gene targeting in *Drosophila* by direct embryo injection with zinc-finger nucleases. *Proc Natl Acad Sci* **105:** 19821–19826.

Bibikova M, Beumer K, Trautman JK, Carroll D. 2003. Enhancing gene targeting with designed zinc finger nucleases. *Science* **300:** 764.

Bottcher R, Hollmann M, Merk K, Nitschko V, Obermaier C, Philippou-Massier J, Wieland I, Gaul U, Forstemann K. 2014. Efficient chromosomal gene modification with CRISPR/cas9 and PCR-based homologous recombination donors in cultured *Drosophila* cells. *Nucleic Acids Res* **42:** e89.

Bozas A, Beumer KJ, Trautman JK, Carroll D. 2009. Genetic analysis of zinc-finger nuclease-induced gene targeting in *Drosophila*. *Genetics* **182:** 641–651.

Carroll D. 2013. Staying on target with CRISPR-Cas. *Nat Biotechnol* **31:** 807–809.

Carroll D. 2014. Genome engineering with targetable nucleases. *Annu Rev Biochem* **83:** 409–439.

Cong L, Ran FA, Cox D, Lin S, Barretto R, Habib N, Hsu PD, Wu X, Jiang W, Marraffini LA, et al. 2013. Multiplex genome engineering using CRISPR/Cas systems. *Science* **339:** 819–823.

Doench JG, Hartenian E, Graham DB, Tothova Z, Hegde M, Smith I, Sullender M, Ebert BL, Xavier RJ, Root DE. 2014. Rational design of highly active sgRNAs for CRISPR-Cas9-mediated gene inactivation. *Nat Biotechnol* **32:** 1262–1267.

Fu Y, Foden JA, Khayter C, Maeder ML, Reyon D, Joung JK, Sander JD. 2013. High-frequency off-target mutagenesis induced by CRISPR–Cas nucleases in human cells. *Nat Biotechnol* **31:** 822–826.

Gratz SJ, Cummings AM, Nguyen JN, Hamm DC, Donohue LK, Harrison MM, Wildonger J, O'Connor-Giles KM. 2013. Genome engineering of *Drosophila* with the CRISPR RNA-guided Cas9 nuclease. *Genetics* **194:** 1029–1035.

Gratz SJ, Ukken FP, Rubinstein CD, Thiede G, Donohue LK, Cummings AM, O'Connor-Giles KM. 2014. Highly specific and efficient CRISPR/Cas9-catalyzed homology-directed repair in *Drosophila*. *Genetics* **196:** 961–971.

Housden BE, Perrimon N. 2016a. Design and generation of donor constructs for genome engineering in *Drosophila*. *Cold Spring Harb Protoc.* doi: 10.1101/pdb.prot090787.

Housden BE, Perrimon N. 2016b. Detection of indel mutations in *Drosophila* by high-resolution melt analysis (HRMA). *Cold Spring Harb Protoc.* doi: 10.1101/pdb.prot090795.

Housden BE, Lin S, Perrimon N. 2014. Cas9-based genome editing in *Drosophila*. *Methods Enzymol* **546:** 415–439.

Housden BE, Hu Y, Perrimon N. 2016. Design and generation of *Drosophila* sgRNA expression constructs. *Cold Spring Harb Protoc.* doi: 10.1101/pdb.prot090779.

Hsu PD, Scott DA, Weinstein JA, Ran FA, Konermann S, Agarwala V, Li Y, Fine EJ, Wu X, Shalem O, et al. 2013. DNA targeting specificity of RNA-guided Cas9 nucleases. *Nat Biotechnol* **31:** 827–832.

Jeggo PA. 1998. DNA breakage and repair. *Adv Genet* **38:** 185–218.

Jiang W, Bikard D, Cox D, Zhang F, Marraffini LA. 2013. RNA-guided editing of bacterial genomes using CRISPR-Cas systems. *Nat Biotechnol* **31:** 233–239.

Jinek M, Chylinski K, Fonfara I, Hauer M, Doudna JA, Charpentier E. 2012. A programmable dual-RNA-guided DNA endonuclease in adaptive bacterial immunity. *Science* **337:** 816–821.

Jinek M, East A, Cheng A, Lin S, Ma E, Doudna J. 2013. RNA-programmed genome editing in human cells. *eLife* **2:** e00471.

Joung JK, Sander JD. 2013. TALENs: A widely applicable technology for targeted genome editing. *Nat Rev Mol Cell Biol* **14:** 49–55.

Kondo S, Ueda R. 2013. Highly improved gene targeting by germline-specific Cas9 expression in *Drosophila*. *Genetics* **195:** 715–721.

Lee H, McManus CJ, Cho DY, Eaton M, Renda F, Somma MP, Cherbas L, May G, Powell S, Zhang D, et al. 2014a. DNA copy number evolution in *Drosophila* cell lines. *Genome Biol* **15:** R70.

Lee JS, Kwak SJ, Kim J, Kim AK, Noh HM, Kim JS, Yu K. 2014b. RNA-guided genome editing in *Drosophila* with the purified Cas9 protein. *G3* **4:** 1291–1295.

Mali P, Aach J, Stranges PB, Esvelt KM, Moosburner M, Kosuri S, Yang L, Church GM. 2013a. CAS9 transcriptional activators for target specificity screening and paired nickases for cooperative genome engineering. *Nat Biotechnol* **31:** 833–838.

Mali P, Yang L, Esvelt KM, Aach J, Guell M, DiCarlo JE, Norville JE, Church GM. 2013b. RNA-guided human genome engineering via Cas9. *Science* **339:** 823–826.

Pattanayak V, Lin S, Guilinger JP, Ma E, Doudna JA, Liu DR. 2013. High-throughput profiling of off-target DNA cleavage reveals RNA-programmed Cas9 nuclease specificity. *Nat Biotechnol* **31:** 839–843.

Ren X, Sun J, Housden BE, Hu Y, Roesel C, Lin S, Liu LP, Yang Z, Mao D, Sun L, et al. 2013. Optimized gene editing technology for *Drosophila melanogaster* using germ line-specific Cas9. *Proc Natl Acad Sci* **110:** 19012–19017.

Ren X, Yang Z, Xu J, Sun J, Mao D, Hu Y, Yang SJ, Qiao HH, Wang X, Hu Q, et al. 2014. Enhanced specificity and efficiency of the CRISPR/Cas9 system with optimized sgRNA parameters in *Drosophila*. *Cell Reports* **9:** 1151–1162.

Tsai SQ, Zheng Z, Nguyen NT, Liebers M, Topkar VV, Thapar V, Wyvekens N, Khayter C, Iafrate AJ, Le LP, et al. 2015. GUIDE-seq enables genome-wide profiling of off-target cleavage by CRISPR–Cas nucleases. *Nat Biotechnol* **33:** 187–197.

van Gent DC, Hoeijmakers JH, Kanaar R. 2001. Chromosomal stability and the DNA double-stranded break connection. *Nat Rev Genet* **2:** 196–206.

Wang T, Wei JJ, Sabatini DM, Lander ES. 2014. Genetic screens in human cells using the CRISPR-Cas9 system. *Science* **343:** 80–84.

Yu Z, Ren M, Wang Z, Zhang B, Rong YS, Jiao R, Gao G. 2013. Highly efficient genome modifications mediated by CRISPR/Cas9 in *Drosophila*. *Genetics* **195:** 289–291.

Yu C, Liu Y, Ma T, Liu K, Xu S, Zhang Y, Liu H, La Russa M, Xie M, Ding S, et al. 2015. Small molecules enhance CRISPR genome editing in pluripotent stem cells. *Cell Stem Cell* **16:** 142–147.

Cite this introduction as *Cold Spring Harb Protoc*; doi:10.1101/pdb.top086843

Protocol 1

Design and Generation of Donor Constructs for Genome Engineering in *Drosophila*

Benjamin E. Housden[1,3] and Norbert Perrimon[1,2,3]

[1]*Department of Genetics, Harvard Medical School, Boston, Massachusetts 02115;* [2]*Howard Hughes Medical Institute, Harvard Medical School, Boston, Massachusetts 02115*

The generation of precise alterations to the genome using CRISPR requires the combination of CRISPR and a donor construct containing homology to the target site. A double-strand break is first generated at the target locus using CRISPR. It is then repaired using the endogenous homologous recombination (HR) pathway. When a donor construct is provided, it can be used as a template for HR repair and can therefore be exploited to introduce alterations in the genomic sequence with single base-pair precision. Here we describe a protocol for the generation of donor constructs using Golden Gate assembly and discuss some key considerations for donor construct design for use in *Drosophila*.

MATERIALS

It is essential that you consult the appropriate Material Safety Data Sheets and your institution's Environmental Health and Safety Office for proper handling of equipment and hazardous materials used in this protocol.

RECIPE: Please see the end of this protocol for recipes indicated by <R>. Additional recipes can be found online at http://cshprotocols.cshlp.org/site/recipes.

Reagents

Agarose

Chemically competent *E. coli* and reagents for transformation

DNA sequencing primers and reagents

- BHF: 5′-GGGAAACGCCTGGTATCTTT-3′
- BHR: 5′-GCATTACGCTGACTTGAC-3′

dNTPs (10 mM) (e.g., New England BioLabs)

Gel purification kit (e.g., QIAGEN)

Genomic DNA template (from the genetic background to be engineered)

Golden Gate assembly reagents

- Adenosine triuphosphate (ATP) (10 mM) (e.g., New England BioLabs)
- Bovine serum albumin (BSA) (10×) (e.g., New England BioLabs)
- NEBuffer 4 (New England BioLabs)

  NEB CutSmart buffer can be used in place of NEBuffer 4 and BSA.

- pBH-donor vector DNA (available from authors on request)

[3]Correspondence: bhousden@genetics.med.harvard.edu; perrimon@receptor.med.harvard.edu

Copyright © Cold Spring Harbor Laboratory Press; all rights reserved

Cite this protocol as *Cold Spring Harb Protoc*; doi:10.1101/pdb.prot090787

T7 ligase (Enzymatics)
Type IIS restriction enzyme (BbsI, BsaI-HF, or BsmBI) (New England BioLabs)

High-fidelity polymerase enzyme and buffer (e.g., Phusion with 5× HF buffer; New England BioLabs)
LB liquid medium <R> with kanamycin for selection

In addition, prepare plates containing LB medium solidified with agar (plus kanamycin).

Oligonucleotide primers to amplify homology arms (10 µM) (designed according to Steps 1–5)
Plasmid DNA miniprep kit (e.g., QIAGEN)
XbaI and XhoI, with CutSmart buffer (New England BioLabs)

Equipment

Incubator at 37°C
NanoDrop or Qubit for DNA quantification
Primer3 (http://bioinfo.ut.ee/primer3) or similar program for primer design
Thermocycler
UV gel imager

METHOD

Designing Homology Arms

The design of a donor construct is highly dependent on the relevant application, although in all cases two homology arms are required. If the intended application is the insertion of sequence into the genome or replacement of endogenous sequence with a transgene, then an insert component is also required in the donor construct. The most important consideration for donor design is the placement of homology arms, which is also dependent on the desired application. For insertions or single base-pair alterations, the two homology arms should be immediately adjacent to one another, whereas for deletions or replacement of endogenous sequence, the homology arms should flank the sequence to be removed. In all cases, the design of homology arms should be paired with the single guide RNA (sgRNA) design, described in Protocol 3: Design and Generation of Drosophila *Single Guide RNA Expression Constructs (Housden et al. 2016). Moreover, in cases where the homology arms are close together, the sgRNA should be as close to the join as possible. Wherever possible, the sgRNA target site should be disrupted by a successful editing event to avoid further modification. In addition, the sgRNA should not target either homology arm. When homology arms are separated (e.g., for deletions), the sgRNA can be placed anywhere within the intervening sequence.*

1. Search for sgRNAs as described in Protocol 3: Design and Generation of *Drosophila* Single Guide RNA Expression Constructs (Housden et al. 2016). Select an sgRNA that targets as close to the intended change in genomic sequence as possible.

   In cases where a large region is to be deleted or replaced, the sgRNA can target anywhere between the two homology arms. In addition, multiple sgRNAs can be expressed simultaneously within this region to increase the editing rate.

2. Design primers to amplify homology arms on either side of the target locus.

   After the optimal sgRNA design is selected and the genomic sequences upstream of as well as downstream from the sgRNA cutting site are retrieved, we recommend using Primer3 (http://bioinfo.ut.ee/primer3) or a similar program for primer design. Polymerase chain reaction (PCR) products should be 500–1200 bp in size. The sequences of both PCR products should be scanned for BbsI, BsaI, and BsmBI restriction enzyme recognition sites to determine the appropriate enzyme for Step 3. If the recognition sites of all the three enzymes are found within the PCR products, the PCR primers must be redesigned.

3. Select a suitable Type IIS restriction enzyme for the Golden Gate reaction: BbsI, BsaI, or BsmBI.

   Ensure that the selected restriction enzyme does not cut within any of the sequences to be assembled. Any suitable backbone vector can be used, but pBH donor was designed to be compatible with multiple different restriction enzymes (BbsI, BsaI, and BsmBI) and contains a ccdB *gene to facilitate the selection of correctly assembled clones. Note that optimal activity of BsmBI occurs at a higher temperature than BsaI or BbsI. This enzyme therefore often results in reduced efficiency using the Golden Gate reaction conditions described below, and BsaI or BbsI should be used preferentially whenever possible.*

4. Add the restriction enzyme binding sequences and overhangs for cloning to the primers designed in Step 2.

 Cite this protocol as *Cold Spring Harb Protoc*; doi:10.1101/pdb.prot090787

The target sequences for BbsI, BsaI, and BsmBI are GAAGACnnXXXX, GGTCTCnXXXX, and CGTCTCnXXXX, respectively. X indicates the four base pairs that will be used for cloning and must be complementary to those on the respective ligation partner. For example, when assembling two homology arms into pBH-donor using BsaI, the sequences added to each primer should be:

Left homology arm forward primer: GGTCTCnGACC-target-specific sequence
Left homology arm reverse primer: GGTCTCnGAAC-target-specific sequence
Right homology arm forward primer: GGTCTCnGTTC-target-specific sequence
Right homology arm reverse primer: GGTCTCnTATA-target-specific sequence

When assembling two homology arms plus an insert into pBH donor, the sequences should be:

Left homology arm forward primer: GGTCTCnGACC-target-specific sequence
Left homology arm reverse primer: GGTCTCnGAAC-target-specific sequence
Insert forward primer: GGTCTCnGTTC-target-specific sequence
Insert reverse primer: GGTCTCnGCCC-target-specific sequence
Right homology arm forward primer: GGTCTCnGGGC-target-specific sequence
Right homology arm reverse primer GGTCTCnTATA-target-specific sequence

5. Synthesize each set of primers and purify by standard desalting.

Generating Donor Constructs

6. Perform PCR to amplify each of the homology fragments (left arm and right arm) and insert, if applicable, using the primer sets from Step 5 as follows.

Reagent	Amount to add
Genomic DNA template	20 ng
5× Phusion HF buffer	10 µL
10 mM dNTPs	1 µL
Forward primer (according to fragment) (10 µM)	1 µL
Reverse primer (according to fragment) (10 µM)	1 µL
Phusion polymerase	0.5 µL

These are standard reaction conditions. Optimization for each fragment may be required.

7. Transfer the samples to a thermocycler and run the following program.

	Temperature	Time
Start	98°C	3 min
29 cycles	98°C	30 sec
	57°C	30 sec
	72°C	30 sec
Hold	10°C	

These are standard cycling conditions. Optimization for each fragment may be required.

8. Purify each fragment product using standard methods for agarose gel purification, and elute in H_2O.
9. Quantify the gel-purified products, and dilute to 20 ng/µL using H_2O.
10. Assemble the Golden Gate reaction as follows.

Reagent	Amount to add
pBH-donor vector	15 ng
Left homology arm	15 ng
Right homology arm	15 ng
Insert fragment (optional)	15 ng
10 mM ATP	1 µL
10× BSA	1 µL
NEB Buffer 4	1 µL
T7 ligase	0.5 µL
Type IIS restriction enzyme	0.5 µL
H_2O	to 10 µL

11. Transfer the samples to a thermocycler and run the following program.

Temperature	Time
37°C	2 min
20°C	3 min
Repeat cycle	14 times
37°C	30 min

12. Transform 5 µL of the reaction products into chemically competent *Escherichia coli* using standard methods. Spread the cells onto plates containing solid LB agar with kanamycin. Incubate overnight at 37°C.

13. Select four individual colonies for culture in LB liquid medium with kanamycin. Prepare plasmid DNA minipreps from each clone using standard methods.

    See Troubleshooting.

14. Assemble the following digestion reaction for each clone.

Reagent	Amount to add
Plasmid miniprep DNA	200 ng
XbaI	0.2 µL
XhoI	0.2 µL
Cutsmart buffer	1 µL
H_2O	to 10 µL

15. Incubate the reactions for 1 h at 37°C.

16. Visualize the digested products on an agarose gel. Calculate the expected band sizes based on the desired product and analyze the gel image to determine whether the correct construct has been produced.

    XbaI and XhoI cut the pBH-donor vector close to the joins with the homology arms. The digested products should therefore include a band at 1624 bp, indicating the presence of the pBH-donor vector, plus additional bands depending on the size and presence of relevant restriction sites in the homology arms or insert sequences.

    See Troubleshooting.

17. Sequence the clones that produced the correct banding patterns in Step 16 using the BHF and BHR sequencing primers.

    For long homology arms, or when an insert sequence is cloned between homology arms, additional sequencing primers will be needed to cover the whole construct.

18. Perform a local sequence alignment to a reference sequence or the experimentally determined sequence of the starting constructs.

TROUBLESHOOTING

Problem (Steps 13 or 16): No colonies are produced from the transformation, or a component is missing from Golden Gate product.

Solution: Golden Gate assembly performs optimally when all components are present in equimolar ratios. For simplicity, we generally calculate the amounts of each component using DNA mass (Step 10). However, depending on the molecular mass of each component, efficiency might be increased by calculating these amounts based on molar ratios. Repeat the assembly process including equimolar amounts of each component.

DISCUSSION

Repair of a CRISPR-induced double-strand break (DSB) using HR is an extremely powerful approach to genome engineering and allows almost any modification to the genomic sequence with single base-

Cite this protocol as *Cold Spring Harb Protoc*; doi:10.1101/pdb.prot090787

pair precision. To achieve this, a donor construct must be provided that contains the modified sequence flanked by regions homologous to the target site, although the specific design will vary depending on the application. In addition to generating a desired modification to the genomic sequence, the use of a donor also allows the inclusion of a selection cassette, greatly easing the identification of successful editing events.

There are two main approaches to donor-based genome engineering that have been used successfully in *Drosophila* (Gratz et al. 2013, 2014; Bassett et al. 2014). A simple option is to generate a single-stranded oligo with short regions of homology (15–50 bp) flanking the DSB site. This option is attractive due to the ease with which these oligos can be generated, but it severely limits the length of sequence that can be inserted into the genome. It is therefore not possible to include a selection cassette using this approach. A second option is to produce a construct containing long homology regions (generally 500–1000 bp) on either side of the DSB site. This allows a much greater length of sequence to be included, but cloning these constructs requires considerably more work than producing a short oligo.

Including a selection cassette or visible marker by using long donor constructs can provide a significant advantage. Small indel mutations, such as those used to generate gene knockouts, generally require indirect molecular screening methods to identify correctly modified animals (Housden et al. 2014). By including a donor construct carrying a visible marker such as *mini-white* or 3XP3-RFP, successfully modified animals can be easily identified in the F_1 generation following injection (Gratz et al. 2014), considerably reducing the effort involved in generating a stable stock. Antibiotic resistance genes could be used in a similar manner in cell culture.

This protocol describes the use of Golden Gate assembly for the production of donor constructs; however, other methods, such as Gibson assembly, are also available (Gibson et al. 2009; Gibson 2011). These approaches are comparable in ease of use and robustness. We generally favor Golden Gate assembly because components can be provided either as linear fragments or intact plasmids. This greatly simplifies the cloning procedure when generating many donors in parallel involving some of the same components. For example, when generating donors for inserting green fluorescent protein (GFP) tags on multiple genes, the GFP can be provided as a plasmid without the need for processing before assembly.

RECIPE

LB Liquid Medium

In 1 L H_2O, dissolve 10 g of Bacto Tryptone, 5 g of yeast extract, and 10 g of NaCl.
Adjust the pH to 7.5 with NaOH. Sterilize by autoclaving.

ACKNOWLEDGMENTS

We thank David Doupé and Stephanie Mohr for helpful discussions during the preparation of this manuscript. Work in the Perrimon Laboratory is supported by grants from the National Institutes of Health and the Howard Hughes Medical Institute.

REFERENCES

Bassett AR, Tibbit C, Ponting CP, Liu JL. 2014. Mutagenesis and homologous recombination in *Drosophila* cell lines using CRISPR/Cas9. *Biol Open* **3:** 42–49.

Gibson DG. 2011. Enzymatic assembly of overlapping DNA fragments. *Methods Enzymol* **498:** 349–361.

Gibson DG, Young L, Chuang RY, Venter JC, Hutchison CA III, Smith HO. 2009. Enzymatic assembly of DNA molecules up to several hundred kilobases. *Nat Methods* **6:** 343–345.

Gratz SJ, Cummings AM, Nguyen JN, Hamm DC, Donohue LK, Harrison MM, Wildonger J, O'Connor-Giles KM. 2013. Genome engineering of *Drosophila* with the CRISPR RNA-guided Cas9 nuclease. *Genetics* **194:** 1029–1035.

Gratz SJ, Ukken FP, Rubinstein CD, Thiede G, Donohue LK, Cummings AM, O'Connor-Giles KM. 2014. Highly specific and efficient CRISPR/Cas9-catalyzed homology-directed repair in *Drosophila*. *Genetics* **196:** 961–971.

Housden BE, Lin S, Perrimon N. 2014. Cas9-based genome editing in *Drosophila*. *Methods Enzymol* **546:** 415–439.

Housden BE, Hu Y, Perrimon N. 2016. Design and generation of *Drosophila* sgRNA expression constructs. *Cold Spring Harb Protoc.* doi: 10.1101/pdb.prot090779.

Protocol 2

Detection of Indel Mutations in *Drosophila* by High-Resolution Melt Analysis (HRMA)

Benjamin E. Housden[1,2] and Norbert Perrimon[1,2,3]

[1]*Department of Genetics, Harvard Medical School, Boston, Massachusetts 02115;* [2]*Howard Hughes Medical Institute, Harvard Medical School, Boston, Massachusetts 02115*

Although CRISPR technology allows specific genome alterations to be created with relative ease, detection of these events can be problematic. For example, CRISPR-induced double-strand breaks are often repaired imprecisely to generate unpredictable short indel mutations. Detection of these events requires the use of molecular screening techniques such as endonuclease assays, restriction profiling, or high-resolution melt analysis (HRMA). Here, we provide detailed protocols for HRMA-based mutation screening in *Drosophila* and analysis of the resulting data using the online tool HRMAnalyzer.

MATERIALS

It is essential that you consult the appropriate Material Safety Data Sheets and your institution's Environmental Health and Safety Office for proper handling of equipment and hazardous materials used in this protocol.

RECIPES: Please see the end of this protocol for recipes indicated by <R>. Additional recipes can be found online at http://cshprotocols.cshlp.org/site/recipes.

Reagents

Agarose
Buffer SB <R>
Chemically competent *E. coli* and reagents for transformation
DNA sequencing primers and reagents
  M13F: 5′-GTAAAACGACGGCCAGT-3′
  M13R: 5′-CAGGAAACAGCTATGACC-3′
dNTPs (10 mM)
Drosophila for HRMA (anesthetized adult flies or cells), with wild-type controls
Gel purification kit (e.g., QIAGEN) (as needed; see Step 18)
High-fidelity polymerase enzyme and buffer (e.g., Phusion with 5× HF buffer; New England BioLabs)
LB liquid medium <R> with ampicillin for selection
  In addition, prepare plates containing LB medium solidified with agar (plus ampicillin).

[3]Correspondence: bhousden@genetics.med.harvard.edu; perrimon@receptor.med.harvard.edu

Copyright © Cold Spring Harbor Laboratory Press; all rights reserved
Cite this protocol as *Cold Spring Harb Protoc*; doi:10.1101/pdb.prot090795

$MgCl_2$ (100 mM)

Oligonucleotide primers for polymerase chain reaction (PCR) and high-resolution melt analysis (HRMA)

We recommend using Primer3 (http://bioinfo.ut.ee/primer3) or a similar program to design primers for PCR and HRMA. Design one set of PCR primers to amplify a 300- to 600-bp fragment surrounding the sgRNA target site (Step 2). Design a second set of HRMA primers to amplify a 45- to 150-bp fragment contained within the sgRNA PCR product (Step 5).

Plasmid DNA miniprep kit (e.g., QIAGEN)

Precision Melt Supermix (Bio-Rad)

Zero Blunt TOPO PCR cloning kit (Invitrogen)

Equipment

Bullet blender or pestles for homogenization in 1.5-mL microcentrifuge tubes (for adult flies)

Centrifuge

Heat blocks or water baths at 50°C and 95°C

HRMAnalyzer (http://www.flyrnai.org/hrma)

Microcentrifuge tubes (1.5 mL)

PCR tubes

RT-PCR (reverse transcription-polymerase chain reaction) machine capable of reading at temperature intervals of 0.1°C

Thermocycler

METHOD

Performing the HRMA Reaction

To generate P-values from HRMA indicating the likelihood of experimental samples containing different sequence versus the controls, at least three control samples (i.e., using wild-type DNA) must be present; however, the quality of output data is improved with additional controls. For the following protocol, we recommend performing at least three control reactions for simple samples (e.g., from diploid flies) and at least eight controls for complex samples (e.g., from mixed cell populations).

1. Prepare genomic DNA from adult flies or cells as follows.

Adult Flies

i. Place an anesthetized fly into a 1.5-mL tube with 50 µL of buffer SB.

ii. Thoroughly homogenize the sample using a bullet blender or pestle.

iii. Centrifuge the sample at maximum speed for 1 min to pellet the remaining tissue fragments. Transfer the supernatant to a PCR tube.

iv. Incubate the supernatant for 1 h at 50°C followed by 10 min at 95°C.

Cells

v. Collect 10,000–100,000 cells in a 1.5-mL microcentrifuge tube. Pellet the cells by centrifugation at 5000*g* for 2 min.

vi. Resuspend the cells in 100 µL of buffer SB.

vii. Incubate the sample for 1 h at 50°C followed by 10 min at 95°C.

2. Assemble the following PCR to amplify the 300- to 600-bp sequence surrounding the sgRNA target locus.

Reagent	Amount to add
Genomic DNA	1 µL
Phusion HF buffer	10 µL
dNTPs (10 mM)	1 µL
Forward PCR primer (10 µM)	1 µL
Reverse PCR primer (10 µM)	1 µL
Phusion polymerase	0.5 µL
$MgCl_2$ (100 mM)	1.25 µL
H_2O	33.25 µL

3. Transfer the reaction to a thermocycler and run the following program.

	Temperature	Time
Start	98°C	3 min
35 cycles	98°C	30 sec
	50°C	30 sec
	72°C	30 sec
Hold	10°C	

This PCR is designed to amplify products with low specificity using a wide range of primer pairs and will likely produce multiple nonspecific bands. In general, it is not necessary to check the product on a gel, as low concentrations are sufficient for the HRMA reaction and nonspecific products are not a concern.

4. Dilute the PCR products 1:10,000 using H_2O.
5. Assemble the HRMA reaction using primers that amplify a 45- to 150-bp fragment contained within the sgRNA PCR product.

Reagent	Amount to add
DNA template from Step 4	1 µL
Precision Melt Supermix	5 µL
Forward HRMA primer (10 µM)	0.3 µL
Reverse HRMA primer (10 µM)	0.3 µL
H_2O	3.4 µL

6. Transfer the reaction to an RT-PCR machine and run the following program.

	Temperature	Time
Start	95°C	3 min
50 cycles	95°C	18 sec
	50°C	30 sec
	Fluorescence read	
1 cycle	95°C	2 min
	25°C	2 min
	4°C	2 min
Melt curve	55°C to 95°C	fluorescence reads every 0.1°C

See Troubleshooting.

Analyzing Data Using HRMAnalyzer

7. Export the HRMA data from the RT-PCR machine with the following format. (See the help file at www.flyrnai.org/HRMAhelp.html.)

Cite this protocol as *Cold Spring Harb Protoc*; doi:10.1101/pdb.prot090795

Column 1	Temperature
Columns 2–End	Fluorescence readings
Row 1	Headings (Temperature; Sample name 1; Sample name 2; etc.)

All control sample names must contain the word "control."

Many RT-PCR machines automatically convert fluorescence readings to −dF/dT. The analysis will not work using this format; thus, be sure to export the raw data (usually called RFU data).

8. Convert the file format to .xlsx or .txt
9. Upload the melt data file to HRMAnalyzer at: http://www.flyrnai.org/hrma (Fig. 1A).
10. Click "Generate Figures."
11. Use the "Data quality cutoff" slider to remove any samples that were not amplified efficiently and click "next."

    Samples not amplified efficiently will have considerably lower fluorescence values than other samples and will often lead to false positive results.

12. Select the upper and lower normalization ranges by dragging the vertical bars or entering values directly into the relevant boxes. Click "Generate Figures" and then "Next" (Fig. 1B).

    The upper normalization range should be positioned to the left of the melting region and the lower range below the melting region (Fig. 1B). Both ranges should cover 3°C–8°C, although this parameter is relatively flexible.

    The "Overlay" option can be used to shift all melt curves along the horizontal axis such that they cross at a point defined by the "Overlay position" slider. This allows comparison of curve shape independent of the melting temperature and can be useful for separation of heterozygous and homozygous samples. It is generally not required and is not recommended for analysis of samples from complex backgrounds, such as in cell culture experiments in which many different sequences may be present.

13. Click "Generate Figures" to obtain a summary graph of the results with hits shown with blue lines and nonhits with black lines (Fig. 1C).

    The clustering option can be used to group samples likely to have the same composition (e.g., carrying the same mutation). However, this option does not perform well on complex samples in which many sequences are present.

14. Click "Download Data" to retrieve data from each step of the analysis process.

    Several files will be downloaded, including the data plotted on the graphs at each analysis step. In addition, a file named "Data_significance.txt" will include a P-value *representing the likelihood of each sample containing sequence different from control samples, and a value labeled "Total area" which can be used to rank the results (high values indicate greater difference from control melt curves).*

    See Troubleshooting.

Analyzing Hits by Sequencing

HRMAnalyzer provides a list of samples likely to have differing sequences from the controls. When working with relatively simple samples, such as those derived from heterozygous mutant flies, this analysis is relatively reliable. However, in cell lines in which many different mutations may be present at potentially low levels, HRMA signals may be weak. It is therefore advisable in these cases to confirm the presence and nature of mutations by sequencing following HRMA.

15. Assemble the following PCR to amplify the region surrounding the mutation.

Reagent	Amount to add
Genomic DNA	1 µL
Phusion HF buffer	10 µL
dNTPs (10 mM)	1 µL
Forward PCR primer (10 µM)	1 µL
Reverse PCR primer (10 µM)	1 µL
Phusion polymerase	0.5 µL
H_2O	34.5 µL

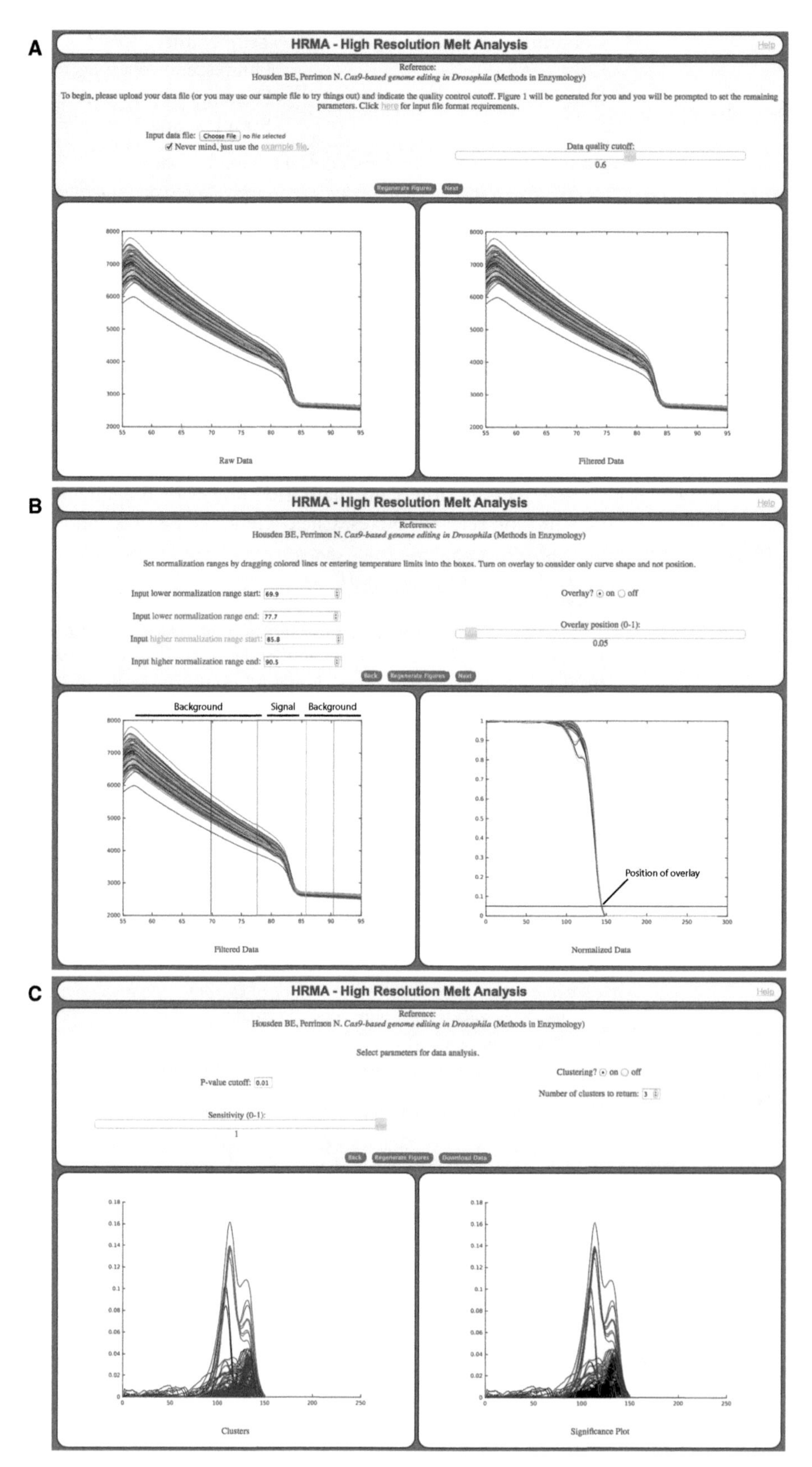

FIGURE 1. HRMAnalyzer interface: Screenshots from the HRMAnalyzer tool. (*A*) Stage 1: Uploading data and removal of poor quality samples from analysis. (*B*) Stage 2: Data normalization. (*C*) Stage 3: Analysis of normalized data.

Cite this protocol as *Cold Spring Harb Protoc*; doi:10.1101/pdb.prot090795

16. Transfer the reaction to a thermocycler and run the following program.

	Temperature	Time
Start	98°C	3 min
30 cycles	98°C	30 sec
	60°C	30 sec
	72°C	30 sec
Hold	10°C	

This PCR may require optimization to produce a specific product.

17. Run a 5-µL aliquot of the PCR product on an agarose gel to check that a band of the correct size was produced.
18. If necessary, gel-purify the correct product and elute in 30 µL of H_2O
19. Assemble a TOPO cloning reaction using the Zero Blunt reagents as follows.

Reagent	Amount to add
Purified PCR product	2 µL
Salt solution	0.5 µL
TOPO vector	0.5 µL

20. Incubate the reaction for 5 min at room temperature.
21. Transform the DNA into chemically competent *E. coli* using standard methods. Spread the cells onto plates containing solid LB agar with ampicillin. Incubate overnight at 37°C.
22. Select individual colonies and culture in 2 mL of LB liquid medium with ampicillin selection overnight at 37°C. Prepare plasmid DNA minipreps from the overnight cultures using standard methods.

    The recommended number of picked colonies is dependent on the source of the samples. When using diploid flies, relatively few colonies can be analyzed to identify all sequences present. Five colonies per sample should be sufficient. When sequencing from cultured cells, many more sequences may be present, so more colonies should be picked and sequenced.

23. Sequence the plasmid DNA constructs using M13F and M13R primers.
24. Perform a local alignment to the experimentally determined starting genomic sequence (if determined), or to a reference genomic sequence, using standard sequence analysis software.

    See Troubleshooting.

TROUBLESHOOTING

Problem (Step 6): The HRMA fluorescence signal is low.
Solution: This result suggests that either the first PCR reaction was inefficient or the HRMA amplification step failed. Run products from the first PCR on a gel and check for a band of the correct size. If a band is not present, optimize the reaction or redesign the primers. If a band is present, redesign primers for the HRM reaction and repeat the assay.

Problem (Step 14): No significance values are produced during analysis.
Solution: This is likely due to insufficient controls. Even if multiple controls are included, some may be filtered due to low data quality. Check that three or more controls are still present in the output data. If not, repeat the assay with additional control samples.

Problem (Step 14): None of the samples are significantly different from controls.
Solution: This may simply mean that no mutations are present. Alternatively, it could be due to noise in the control samples. Inspect the data in file "Data_normalized.txt" for control samples to check

that the curves are very similar. If they are not, repeat the assay using fresh control DNA to improve quality.

Problem (Step 24): Significantly different samples do not contain mutations when analyzed by sequencing.

Solution: This can occur when HRMA primers produce nonspecific bands. Run HRMA reaction products on a gel to ensure that a single band is produced.

DISCUSSION

The detection of indel mutations induced using CRISPR requires the use of molecular screening techniques, as no markers are introduced to indicate the mutation event. Several methods are available to screen for indel mutations. For example, where an indel mutation is likely to disrupt a restriction site, restriction profiling can be used to determine the presence and frequency of mutations (Nekrasov et al. 2013; Liang et al. 2014). Alternatively, endonuclease assays have been developed to detect the heteroduplexes formed when mutant DNA sequences are mixed with wild-type (Cho et al. 2013; Cong et al. 2013; Vouillot et al. 2015). Finally, HRMA can be used to detect the presence of sequence alterations such as indel mutations (Dahlem et al. 2012; Xing et al. 2012; Bassett et al. 2013). One advantage of HRMA over other assays is that many samples can be analyzed in parallel with little increase in workload. In addition, HRMA is extremely sensitive and can detect rare mutations or complex mixtures of mutations that may arise from genome editing in cultured cells and might be missed using endonuclease assays. Methods for application of these approaches in *Drosophila* have been described elsewhere (Bassett and Liu 2014; Beumer and Carroll 2014; Housden et al. 2014). Here we have focused on HRMA and data analysis using the HRMAnalyzer tool, as this technique can initially appear complex and is covered less extensively in previous literature.

RECIPES

LB Liquid Medium

In 1 L H_2O, dissolve 10 g of Bacto Tryptone, 5 g of yeast extract, and 10 g of NaCl. Adjust the pH to 7.5 with NaOH. Sterilize by autoclaving.

Buffer SB

10 mM Tris-HCl (pH 8.2)
1 mM EDTA
25 mM NaCl
400 µg/mL Proteinase K (added fresh before use)

ACKNOWLEDGMENTS

We thank David Doupé and Stephanie Mohr for helpful discussions during the preparation of this manuscript. Work in the Perrimon Laboratory is supported by grants from the National Institutes of Health and the Howard Hughes Medical Institute.

REFERENCES

Bassett A, Liu JL. 2014. CRISPR/Cas9 mediated genome engineering in *Drosophila*. *Methods* **69:** 128–136.

Bassett AR, Tibbit C, Ponting CP, Liu JL. 2013. Highly efficient targeted mutagenesis of *Drosophila* with the CRISPR/Cas9 system. *Cell Rep* **4:** 220–228.

Beumer KJ, Carroll D. 2014. Targeted genome engineering techniques in *Drosophila*. *Methods* **68:** 29–37.

Cho SW, Kim S, Kim JM, Kim JS. 2013. Targeted genome engineering in human cells with the Cas9 RNA-guided endonuclease. *Nat Biotechnol* **31:** 230–232.

Cite this protocol as *Cold Spring Harb Protoc*; doi:10.1101/pdb.prot090795

Cong L, Ran FA, Cox D, Lin S, Barretto R, Habib N, Hsu PD, Wu X, Jiang W, Marraffini LA, et al. 2013. Multiplex genome engineering using CRISPR/Cas systems. *Science* **339:** 819–823.

Dahlem TJ, Hoshijima K, Jurynec MJ, Gunther D, Starker CG, Locke AS, Weis AM, Voytas DF, Grunwald DJ. 2012. Simple methods for generating and detecting locus-specific mutations induced with TALENs in the zebrafish genome. *PLoS Genet* **8:** e1002861.

Housden BE, Lin S, Perrimon N. 2014. Cas9-based genome editing in *Drosophila. Methods Enzymol* **546:** 415–439.

Liang Z, Zhang K, Chen K, Gao C. 2014. Targeted mutagenesis in *Zea mays* using TALENs and the CRISPR/Cas system. *J Genet Genomics* **41:** 63–68.

Nekrasov V, Staskawicz B, Weigel D, Jones JD, Kamoun S. 2013. Targeted mutagenesis in the model plant *Nicotiana benthamiana* using Cas9 RNA-guided endonuclease. *Nat Biotechnol* **31:** 691–693.

Vouillot L, Thelie A, Pollet N. 2015. Comparison of T7E1 and surveyor mismatch cleavage assays to detect mutations triggered by engineered nucleases. *G3 (Bethesda)* **5:** 407–415.

Xing L, Hoshijima K, Grunwald DJ, Fujimoto E, Quist TS, Sneddon J, Chien CB, Stevenson TJ, Bonkowsky JL. 2012. Zebrafish foxP2 zinc finger nuclease mutant has normal axon pathfinding. *PLoS One* **7:** e43968.

Protocol 3

Design and Generation of *Drosophila* Single Guide RNA Expression Constructs

Benjamin E. Housden,[1,3] Yanhui Hu,[1] and Norbert Perrimon[1,2,3]

[1]*Department of Genetics, Harvard Medical School, Boston, Massachusetts 02115;* [2]*Howard Hughes Medical Institute, Harvard Medical School, Boston, Massachusetts 02115*

The recent advances in CRISPR-based genome engineering have enabled a plethora of new experiments to study a wide range of biological questions. The major attraction of this system over previous methods is its high efficiency and simplicity of use. For example, whereas previous genome engineering technologies required the generation of new proteins to target each new locus, CRISPR requires only the expression of a different single guide RNA (sgRNA). This sgRNA binds to the Cas9 endonuclease protein and directs the generation of a double-strand break to a highly specific genomic site determined by the sgRNA sequence. In addition, the relative simplicity of the *Drosophila* genome is a particular advantage, as possible sgRNA off-target sites can easily be avoided. Here, we provide a step-by-step protocol for designing sgRNA target sites using the *Drosophila* RNAi Screening Center (DRSC) Find CRISPRs tool (version 2). We also describe the generation of sgRNA expression plasmids for the use in cultured *Drosophila* cells or in vivo. Finally, we discuss specific design requirements for various genome engineering applications.

MATERIALS

It is essential that you consult the appropriate Material Safety Data Sheets and your institution's Environmental Health and Safety Office for proper handling of equipment and hazardous materials used in this protocol.

RECIPE: Please see the end of this protocol for recipes indicated by <R>. Additional recipes can be found online at http://cshprotocols.cshlp.org/site/recipes.

Reagents

BbsI restriction enzyme (Thermo Scientific)
Chemically competent *E. coli* and reagents for transformation
DNA sequencing primer (5′-CAATAGGACACTTTGATTC-3′) and reagents
FastAP enzyme and 10× FastDigest buffer (Thermo Scientific)
LB liquid medium <R> with ampicillin for selection

In addition, prepare plates containing LB medium solidified with agar (plus ampicillin).

PCR purification kit (e.g., QIAGEN)
Plasmid DNA miniprep kit (e.g., QIAGEN)
sgRNA cloning vector (pl100 or pl18 [Ren et al. 2013; Housden et al. 2015]) (available from authors on request)

[3]Correspondence: bhousden@genetics.med.harvard.edu; perrimon@receptor.med.harvard.edu

Copyright © Cold Spring Harbor Laboratory Press; all rights reserved
Cite this protocol as *Cold Spring Harb Protoc*; doi:10.1101/pdb.prot090779

sgRNA oligonucleotides (100 μM) (designed according to Steps 8–12)
T4 polynucleotide kinase (T4 PNK) and 10× T4 ligation buffer (New England BioLabs)
T7 ligase and 2× Quick ligase buffer (Enzymatics)

Equipment

Incubator at 37°C
sgRNA selection tool (DRSC Find CRISPRs [version 2]; www.flyrnai.org/crispr2)

Because of the abundance of possible target sites in the Drosophila *genome, it is often relatively simple to identify sgRNA sequences close to the locus of interest. However, both efficiency and specificity vary greatly between different sgRNAs; therefore, careful target site selection is critical to successful genome modification. Various tools exist for the consideration of variables such as off-target (OT) prediction and mutation efficiency to aid in the selection of the optimal sgRNA for any given application. DRSC Find CRISPRs allows visualization of all possible sgRNA target sites on both strands relevant to any gene or any specific location throughout the* Drosophila *genome in the context of a genome browser (Ren et al. 2013; Housden et al. 2014). Each sgRNA is annotated with predicted OT sites, a score that predicts mutation efficiency, and restriction enzyme recognition site(s) that might be impacted by DSB. The user has the option to select different thresholds for predicted OT potential as well as apply filters based on the relationship of the OT site to gene annotations. For example, a user can opt to allow sgRNAs with OT predictions if they are not located in the coding DNA sequence (CDS) region of any other gene. Other available tools, each with different criteria for sgRNA target annotation and OT prediction, include CRISPR Optimal Target Finder (http://tools.flycrispr.molbio.wisc.edu/targetFinder/) (Gratz et al. 2014), E-CRISP (*http://www.e-crisp.org/E-CRISP/designcrispr.html*) (Heigwer et al. 2014), and CHOPCHOP (https://chopchop.rc.fas.harvard.edu/index.php) (Montague et al. 2014).*

Thermocycler

METHOD

Designing sgRNAs Using DRSC Find CRISPRs

1. Load the home page at http://www.flyrnai.org/crispr2 (Fig. 1A).
2. Enter the FlyBase gene ID, CG number, gene symbol, or chromosome location of the desired editing event (Box 1).
3. Select the OT requirements and features of the targeted sequence (e.g., CDS, untranslated region [UTR], intron, intergenic) (Box 2).
4. Select PAM sequences to include for use with Cas9 derived from *Streptococcus pyogenes* (Box 3).

   Note that NGG results in considerably higher efficiency than NAG.

5. Select OT stringency (Box 4).

   The least stringent option considers sites with up to three mismatches as OTs. The most stringent option considers sites with up to five mismatches as OTs.

6. Click "Submit" to view all sgRNA target sites in the context of a genome browser. Identify the region of interest.

   Note that the first CDS exon can be either to the left or to the right, depending on the strand.

   See Troubleshooting.

7. Click each sgRNA site to view detailed information such as OT annotation and efficiency score (Fig. 1B). Select the optimal sgRNA designs according to the guidelines provided in Figure 1C.

   See Discussion.

Cloning sgRNAs

Designing Oligonucleotides

8. Copy the selected sgRNA target sequence from the DRSC Find CRISPRs tool.
9. Remove the NGG PAM from the 3′ end to obtain the sense sgRNA oligo sequence.
10. To obtain the antisense oligo sequence, find the reverse complement of the sense oligo.

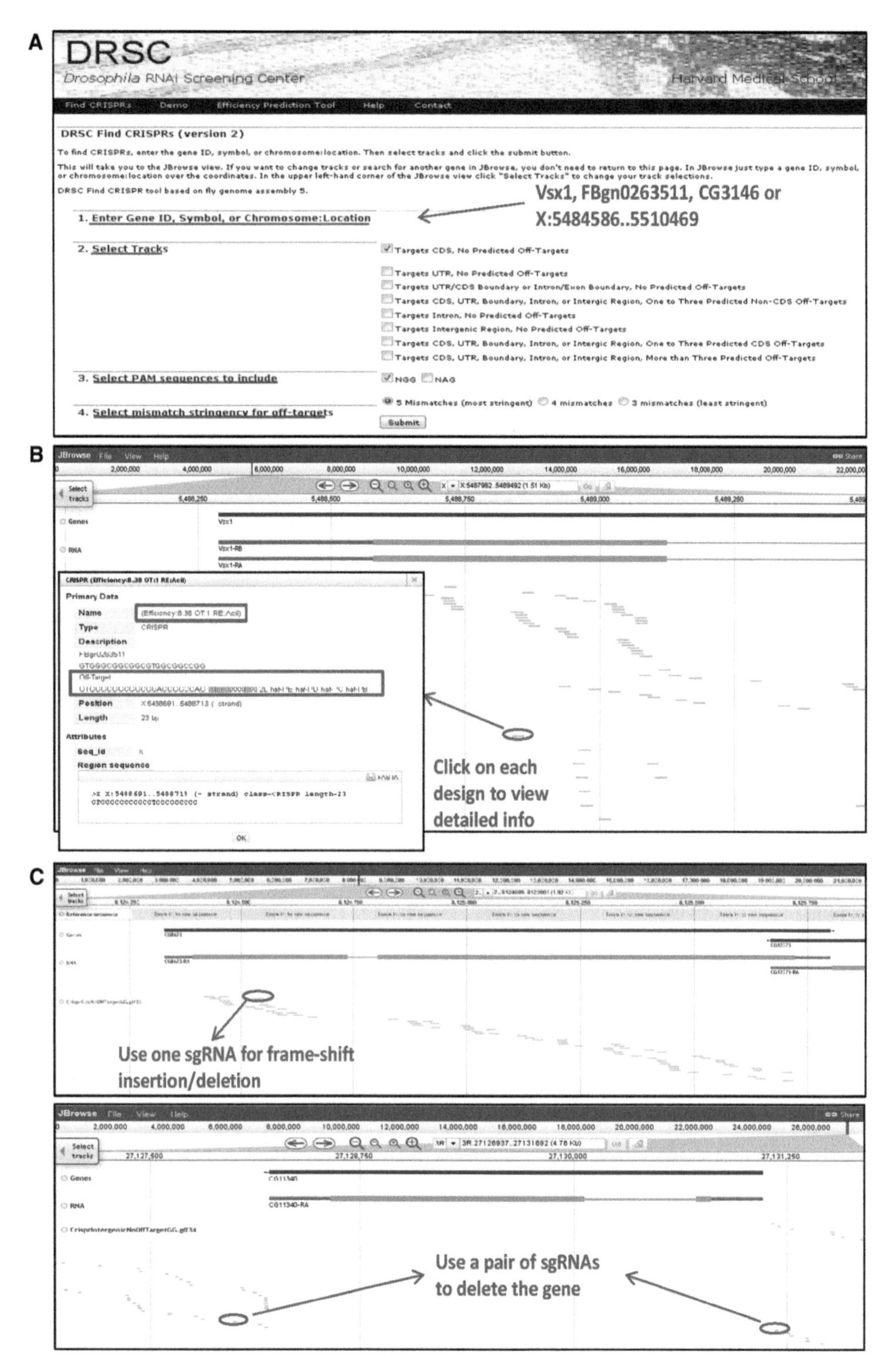

FIGURE 1. The DRSC Find CRISPRs tool (version 2). (*A*) At the query page, the user may enter gene symbol, CG number, FlyBase gene ID, or genome coordinates, and then select criteria for the target region (e.g., CDS, UTR, or intron) and OT stringency. (*B*) The user can view all relevant CRISPR target sites in the context of a genome browser. To view detailed information including sequence, the user clicks on an sgRNA (green bar) in the desired target region. (*C*) Examples of two frequently used strategies for selecting sgRNAs for gene knockout. Note that for some genes, the 5′ region of the gene will be on the right (i.e., on the opposite strand relative to the display).

 Cite this protocol as *Cold Spring Harb Protoc*; doi:10.1101/pdb.prot090779

11. Add the following cloning sequences to the sense and antisense oligos.

Cloning vector	Sequences to add	
For pl100	sense: "CTTCG-20 bp"	antisense: "AAAC-20 bp-C"
For pl18	sense: "GTTCG-20 bp"	antisense: "AAAC-20 bp-C"

Both pl100 and pl18 express the sgRNA from a U6 promoter. This requires that the expressed sequence start with G. We therefore add G to the start of every sgRNA, although this is not necessary when the first position of the sgRNA target is already G.

12. Synthesize the sense and antisense oligos and purify by standard desalting. Resuspend the oligos in H_2O to a final concentration of 100 µM.

Annealing sgRNA Oligos

13. Assemble the following reaction mixture.

Reagent	Amount to add
Sense sgRNA oligo (100 µM)	1 µL
Antisense sgRNA oligo (100 µM)	1 µL
10× T4 ligation buffer	1 µL
T4 PNK	0.5 µL
H_2O	6.5 µL

14. Transfer the mixture to a thermocycler and run the following program.

Temperature	Time
37°C	30 min
95°C	5 min
Ramp from 95°C to 25°C	5°C/min

15. Dilute the annealed oligos 200-fold using H_2O.

The final oligo concentration is 50 nM.

Linearizing the Cloning Vector

16. Assemble the following digestion reaction to linearize the selected cloning vector.

Reagent	Amount to add
Vector DNA (pl100 or pl18)	1 µg
10× FastDigest buffer	2 µL
FastAP enzyme	1 µL
BbsI	1 µL
H_2O	to 20 µL

17. Incubate the reaction mixture for 30 min at 37°C.
18. Purify the reaction products using a PCR purification kit following the manufacturer's instructions.
19. Dilute the linearized plasmid to 10 ng/µL using H_2O.

A large preparation of linearized plasmid can be produced and stored at −20°C for later use.

Ligating the sgRNA Fragment into the Linearized Vector

When performing this protocol for the first time, we recommend including a negative control in which the annealed oligos are omitted. This sample should not produce colonies if the vector is correctly linearized.

20. Assemble the following ligation reaction mixture.

Reagent	Amount to add
Linearized plasmid (10 ng) from Step 19	1 µL
Annealed oligos (50 nM) from Step 15	1 µL
2× Quick ligase buffer	5 µL
T7 ligase	0.5 µL
H_2O	2.5 µL

21. Incubate the ligation mixture for 5 min at room temperature.
22. Transform 2 µL of ligation mixture into chemically competent *E. coli* using standard methods. Spread the cells onto plates containing solid LB agar with ampicillin. Incubate overnight at 37°C.
23. Select two individual colonies for culture in LB liquid medium with ampicillin. Prepare plasmid DNA minipreps from each clone using standard methods.
24. Sequence the miniprep DNA using the sequencing primer (5′-CAATAGGACACTTTGATTC-3′) to check for successful cloning of a single sgRNA sequence.
    See Troubleshooting.

TROUBLESHOOTING

Problem (Step 6): There are no suitable sgRNA target sites.
Solution: Although sgRNA target sites are common throughout the genome, there are regions in which suitable sequences are sparse. For example, PAM sequences may be less common in AT-rich regions such as introns. To increase the number of target sites displayed by the DRSC Find CRISPRs tool, reduce the OT stringency threshold, select more tracks with different OT features, or select NAG instead of NGG for the PAM sequence (Fig. 1A).

Problem (Step 24): No oligo inserts are detected, or there are multiple oligo inserts present in the cloning vector.
Solution: The method described is generally extremely robust and only one or two clones need to be sequenced to find a correctly cloned construct. If no insert is detected, it is likely that the vector linearization was incomplete. This can be checked by performing a negative control ligation omitting the annealed oligos. A completely linearized vector should produce zero colonies under these conditions. If multiple inserts are detected, increase the ratio of oligo dilution to 1:500 in Step 15.

DISCUSSION

The above protocol provides a general idea of how to use DRSC Find CRISPRs (version 2) to generate sgRNA expression plasmids. However, the specific positioning and nature of the sgRNA will vary depending on the desired application. For example, to generate a frameshift gene knockout, a single sgRNA targeting downstream and close to the start of the coding sequence should be designed. In contrast, to delete a defined locus, two sgRNAs flanking that region or one or two sgRNAs in combination with a donor construct should be used. The specific requirements for some common applications are illustrated in Figure 1C.

One major concern associated with the use of CRISPR in mammalian systems has been the occurrence of OT effects. Indeed, a recent report indicated that OT mutations occur at high efficiency and at multiple loci (Tsai et al. 2015). In addition, many of the OT sites identified in the study were not predicted by current algorithms, making the specificity of any given sgRNA unknown.

In contrast, OT effects appear to be much less of a concern in *Drosophila*, and although several groups have reported high-efficiency on-target mutation, OTs have not yet been observed. However, unbiased OT detection methods such as GUIDE-seq (Tsai et al. 2015) have not yet been used in

Cite this protocol as *Cold Spring Harb Protoc*; doi:10.1101/pdb.prot090779

Drosophila, so it is possible that OT mutations remain undetected. From current prediction algorithms, it appears that 97% of genes can be targeted without OT effects (Ren et al. 2013), suggesting that with careful sgRNA design, OTs are not a major concern in *Drosophila*.

Several modifications to the CRISPR system have been developed to reduce the occurrence of OT mutations. For example, Cas9 can be converted to a nickase to generate single-strand breaks by mutating one of the two endonuclease domains. Single-strand breaks are considerably less likely to lead to indel mutations than double-strand breaks (DSBs) but can still be repaired by homologous recombination (HR) (Davis and Maizels 2014; Rong et al. 2014). This version of the system can therefore be used to make targeted changes with reduced OT mutation frequency. However, Cas9 nickase appears to be considerably less efficient than wild-type Cas9 at inducing HR events, and OT indels may still occur at a low level (Ren et al. 2014). Despite these disadvantages, this is a useful approach where high specificity is important.

A modification of the nickase approach is to use two nickase Cas9 constructs targeting sites in close proximity (Mali et al. 2013; Ran et al. 2013, 2014; Cho et al. 2014). This leads to a DSB only at loci bound by both sgRNAs, whereas OT sites will be cleaved only on a single DNA strand. This likely biases the generation of indel mutations toward the on-target site, but similar to the single nickase approach, efficiency is reduced compared to wild-type Cas9 (Ran et al. 2013). In addition, although the use of two sgRNAs may decrease OT mutation at any given site, it may increase the number of OT sites at which single-strand breaks are generated.

A common application of CRISPR is the induction of frameshift mutations to generate gene knockouts. For this to be successful, DSBs must be induced efficiently but the resulting mutation must also generate a frameshift. Although the repair process often appears to be random, microhomology in the surrounding sequence leads to bias in the resulting repaired sequence (Bae et al. 2014). These effects can be predicted by considering the extent of microhomology and distance from the DSB site. A tool was recently developed to predict the likelihood of generating a frameshift mutation (Bae et al. 2014). One future application for which this may be particularly useful is tissue-specific mutagenesis. It was recently shown that CRISPR can be implemented in defined *Drosophila* tissues to generate inducible mutations (Xue et al. 2014). Note that the DRSC Find CRISPRs tool described above includes predicted frameshift scores for all sgRNAs in the *Drosophila* genome as well as a stand-alone page for users to calculate frameshift scores for their own designs (http://www.flyrnai.org/evaluateCrispr/).

RECIPE

LB Liquid Medium

In 1 L H_2O, dissolve 10 g of Bacto Tryptone, 5 g of yeast extract, and 10 g of NaCl. Adjust the pH to 7.5 with NaOH. Sterilize by autoclaving.

ACKNOWLEDGMENTS

We thank David Doupé and Stephanie Mohr for helpful discussions during the preparation of this manuscript. Work in the Perrimon Laboratory is supported by grants from the National Institutes of Health and the Howard Hughes Medical Institute.

REFERENCES

Bae S, Kweon J, Kim HS, Kim JS. 2014. Microhomology-based choice of Cas9 nuclease target sites. *Nat Methods* **11:** 705–706.

Cho SW, Kim S, Kim Y, Kweon J, Kim HS, Bae S, Kim JS. 2014. Analysis of off-target effects of CRISPR/Cas-derived RNA-guided endonucleases and nickases. *Genome Res* **24:** 132–141.

Davis L, Maizels N. 2014. Homology-directed repair of DNA nicks via pathways distinct from canonical double-strand break repair. *Proc Natl Acad Sci* **111:** E924–E932.

Gratz SJ, Ukken FP, Rubinstein CD, Thiede G, Donohue LK, Cummings AM, O'Connor-Giles KM. 2014. Highly specific and efficient CRISPR/

Cas9-catalyzed homology-directed repair in *Drosophila. Genetics* **196:** 961–971.
Heigwer F, Kerr G, Boutros M. 2014. E-CRISP: Fast CRISPR target site identification. *Nat Methods* **11:** 122–123.
Housden BE, Lin S, Perrimon N. 2014. Cas9-based genome editing in *Drosophila. Methods Enzymol* **546:** 415–439.
Housden BE, Valvezan AJ, Kelley C, Sopko R, Hu Y, Roesel C, Lin S, Buckner M, Tao R, Yilmazel B, et al. 2015. Identification of potential drug targets for tuberous sclerosis complex by synthetic screens combining CRISPR-based knockouts with RNAi. *Sci Signal* **8:** RS9.
Mali P, Aach J, Stranges PB, Esvelt KM, Moosburner M, Kosuri S, Yang L, Church GM. 2013. CAS9 transcriptional activators for target specificity screening and paired nickases for cooperative genome engineering. *Nat Biotechnol* **31:** 833–838.
Montague TG, Cruz JM, Gagnon JA, Church GM, Valen E. 2014. CHOPCHOP: A CRISPR/Cas9 and TALEN web tool for genome editing. *Nucleic Acids Res* **42:** W401–W407.
Ran FA, Hsu PD, Lin CY, Gootenberg JS, Konermann S, Trevino AE, Scott DA, Inoue A, Matoba S, Zhang Y, et al. 2013. Double nicking by RNA-guided CRISPR Cas9 for enhanced genome editing specificity. *Cell* **154:** 1380–1389.
Ren X, Sun J, Housden BE, Hu Y, Roesel C, Lin S, Liu LP, Yang Z, Mao D, Sun L, et al. 2013. Optimized gene editing technology for *Drosophila melanogaster* using germ line-specific Cas9. *Proc Natl Acad Sci* **110:** 19012–19017.
Ren X, Yang Z, Mao D, Chang Z, Qiao HH, Wang X, Sun J, Hu Q, Cui Y, Liu LP, et al. 2014. Performance of the Cas9 nickase system in *Drosophila melanogaster. G3 (Bethesda)* **4:** 1955–1962.
Rong Z, Zhu S, Xu Y, Fu X. 2014. Homologous recombination in human embryonic stem cells using CRISPR/Cas9 nickase and a long DNA donor template. *Protein Cell* **5:** 258–260.
Tsai SQ, Zheng Z, Nguyen NT, Liebers M, Topkar VV, Thapar V, Wyvekens N, Khayter C, Iafrate AJ, Le LP, et al. 2015. GUIDE-seq enables genome-wide profiling of off-target cleavage by CRISPR–Cas nucleases. *Nat Biotechnol* **33:** 187–197.
Xue Z, Wu M, Wen K, Ren M, Long L, Zhang X, Gao G. 2014. CRISPR/Cas9 mediates efficient conditional mutagenesis in *Drosophila. G3 (Bethesda)* **4:** 2167–2173.

Cite this protocol as *Cold Spring Harb Protoc*; doi:10.1101/pdb.prot090779

CHAPTER 9

Optimization Strategies for the CRISPR–Cas9 Genome-Editing System

Charles E. Vejnar,[1,4] Miguel A. Moreno-Mateos,[1,4] Daniel Cifuentes,[1,3] Ariel A. Bazzini,[1] and Antonio J. Giraldez[1,2,5]

[1]*Department of Genetics, Yale University School of Medicine, New Haven, Connecticut 06510;* [2]*Yale Stem Cell Center, Yale University School of Medicine, New Haven, Connecticut 06520*

The CRISPR–Cas9 system uncovered in bacteria has emerged as a powerful genome-editing technology in eukaryotic cells. It consists of two components—a single guide RNA (sgRNA) that directs the Cas9 endonuclease to a complementary DNA target site. Efficient targeting of individual genes requires highly active sgRNAs. Recent efforts have made significant progress in understanding the sequence features that increase sgRNA activity. In this introduction, we highlight advancements in the field of CRISPR–Cas9 targeting and discuss our web tool CRISPRscan, which predicts the targeting activity of sgRNAs and improves the efficiency of the CRISPR–Cas9 system for in vivo genome engineering.

OPTIMIZING CRISPR–Cas9 TARGETING EFFICIENCY

The CRISPR–Cas9 system has facilitated rapid in vivo reverse genetics studies across multiple systems (Friedland et al. 2013; Hwang et al. 2013; Wang et al. 2013; Bassett and Liu 2014), but the optimal design of single guide RNAs (sgRNAs) is essential to maximize the efficiency of the system. Using a cell proliferation screen, Wang et al. (2014) first suggested that GC-rich sgRNAs improved targeting efficiency, whereas poly(U) stretches close to the protospacer-adjacent motif (PAM) sequence (Fig. 1) were associated with sgRNAs of lower efficiency. This effect was later attributed to premature termination during sgRNA transcription, given the resemblance of the poly(U) stretches to the RNA polymerase III termination sequence (Wu et al. 2014).

Using a loss-of-function screen targeting nine genes coding for cell-surface proteins, Doench et al. (2014) identified nucleotide biases affecting the activity of thousands of sgRNAs in mammalian cell lines. By analyzing the targeted sequence and the flanking nucleotides (Fig. 1), they observed a significant guanine enrichment 1 nt upstream of the PAM sequence. This strong bias was also observed in vivo (Gagnon et al. 2014; Farboud and Meyer 2015). Combining multiple CRISPR–Cas9 screens, Xu et al. (2015) proposed an improved model for the sgRNA design. They observed a large overlap of the nucleotide biases among these screens and proposed a model to predict sgRNA efficiency based on the consensus features. Recently, Chari et al. (2015) applied a high-throughput sequencing approach to measure sgRNA activity in a large-scale screen. In contrast with the Doench et al. (2014) approach, they used a shorter activity time of CRISPR–Cas9 (72 h vs. 2 wk) and a non-phenotype-based readout

[3]Present address: Department of Biochemistry, Boston University School of Medicine, Boston, Massachusetts 02118

[4]These authors contributed equally to this work.

[5]Correspondence: antonio.giraldez@yale.edu

Copyright © Cold Spring Harbor Laboratory Press; all rights reserved

Cite this introduction as *Cold Spring Harb Protoc;* doi:10.1101/pdb.top090894

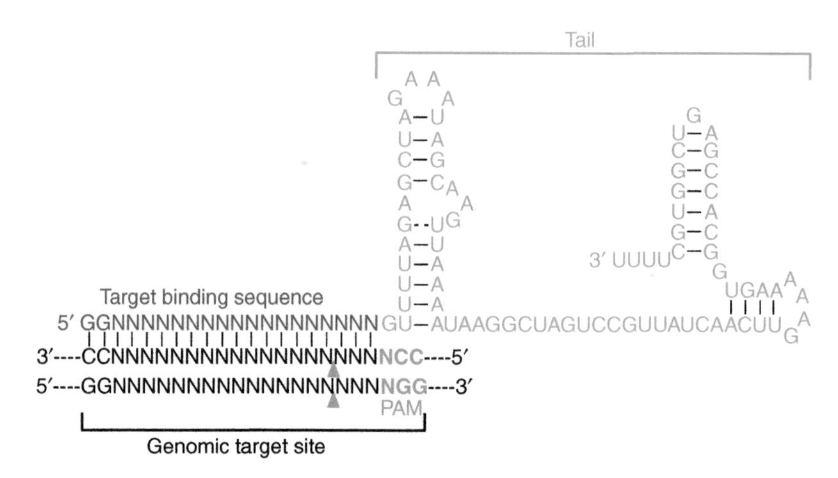

FIGURE 1. sgRNA–target site interaction. An sgRNA (target-binding sequence in red; tail in light blue) binds to its genomic target site (black), which is adjacent to the PAM sequence 5′-NGG (green). Cleavage is predicted to occur at the sites indicated by orange triangles. (Adapted by permission from Macmillan Publishers Ltd: *Nature Methods* [Moreno-Mateos et al. 2015], © 2015.)

(DNA sequencing vs. protein detection). Although the same G-rich bias upstream of the PAM sequence was reported, there was a weak correlation between the most efficient sgRNAs predicted by each study. This might underline the specificity of Cas9 activity in different systems and approaches, although some rules, such as the G-rich upstream of the PAM sequence, are widely applicable.

By testing 1280 sgRNAs in vivo using zebrafish embryos as a model system, we recapitulated the biases of the CRISPR–Cas9 system described above and uncovered specific features of efficient sgRNAs. We integrated these features into the CRISPRscan model, which we also validated in *Xenopus* (Moreno-Mateos et al. 2015). In addition, because we directly provided in vitro–transcribed sgRNAs, we were able to identify features associated with sgRNA stability that correlate with stronger activities. Recent studies of chemically modified sgRNAs have shown that more stable sgRNAs are more active in primary human cells (Hendel et al. 2015). These results highlight that the stability of the sgRNA molecule influences sgRNA activity when the sgRNA is exogenously provided rather than endogenously transcribed. Alternatively, differential stability influencing sgRNA activity may be controlled through delivery of preassembled Cas9–sgRNA ribonucleoprotein complexes (Gagnon et al. 2014; Kim et al. 2014).

EXPANDING THE TARGETING REPERTOIRE

Apart from the predicted targeting efficiency, precise mutagenesis is limited by the frequency of the specific PAM in the targeted genome and the sequence constraints to produce the sgRNA. The adapted CRISPR–Cas9 system used most extensively today for gene targeting is based on the Type II CRISPR–Cas9 system from *Streptococcus pyogenes* (Cong et al. 2013; Jinek et al. 2013; Mali et al. 2013). The Type II endonuclease Cas9 from these bacteria recognizes a PAM sequence next to the target consisting of 5′-NGG (Fig. 1). In addition, sgRNA sequences are further limited by their transcription requirements: sgRNAs require either (i) a G at the 5′ end of the molecule when using ex vivo RNA polymerase III–based systems or (ii) GA/GG when produced in vitro with SP6 or T7/T3 promoters. These restrictions limit the number of potential targets in the genome to 5′-G[N_{20}]GG in the case of RNA polymerase III–based systems and to 5′-G[G/A][N_{19}]GG for in vitro–transcribed sgRNAs.

To circumvent this constraint, various approaches have been recently developed. First, Cas9 orthologs from other bacteria (*Neisseria meningitidis*, *Streptococcus thermophilus*, and *Staphylococcus aureus*) have also been shown to target eukaryotic genomes (Esvelt et al. 2013; Hou et al. 2013; Ran et al. 2015). However, these orthologous Cas9s have PAM sequences longer or similar to those of

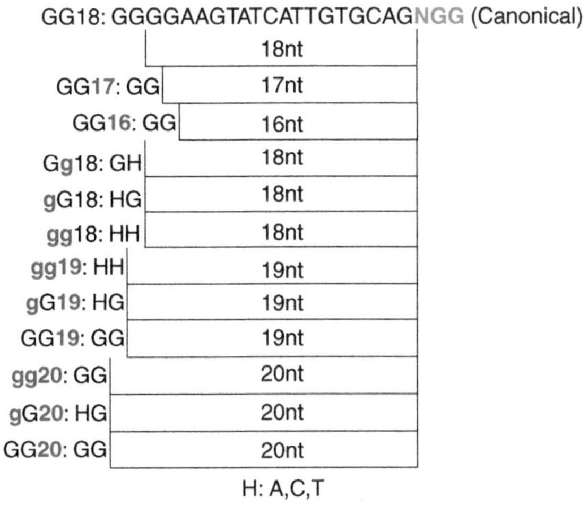

GG18: GGGGAAGTATCATTGTGCAGNGG (Canonical)
18nt
GG17: GG 17nt
GG16: GG 16nt
Gg18: GH 18nt
gG18: HG 18nt
gg18: HH 18nt
gg19: HH 19nt
gG19: HG 19nt
GG19: GG 19nt
gg20: GG 20nt
gG20: HG 20nt
GG20: GG 20nt
H: A,C,T

FIGURE 2. The 11 classes of alternative sgRNA targets analyzed in Moreno-Mateos et al. (2015). The PAM sequence and the alternative features are highlighted in green and in red, respectively. Mismatches between the sgRNAs and the targets are indicated by lowercase letters. (Adapted by permission from Macmillan Publishers Ltd: *Nature Methods* [Moreno-Mateos et al. 2015], © 2015.)

S. pyogenes (e.g., *S. aureus* PAM: 5′-NNGRRT), which do not dramatically increase the number of targets in a genome. To overcome this limitation, Kleinstiver et al. (2015) engineered the *S. pyogenes* Cas9 to recognize different PAM sequences (5′-NGA and 5′-NGCG), doubling the number of the targets in the human genome. More recently, a new endonuclease named Cpf1 was characterized (Zetsche et al. 2015), providing a significant increase in the number of genomic targets due to a drastically different PAM sequence (5′-TTTN).

As a complementary approach, we performed a large-scale analysis in which we sought efficient sgRNAs that target sequences other than the canonical G[G/A][N_{19}]GG (Moreno-Mateos et al. 2015). We analyzed the activity of 11 alternative targeting formulations in zebrafish embryos, varying the lengths of the sgRNAs, and introducing mismatches to the first two nucleotides of the target site (Fig. 2). We found that sgRNAs truncated by 1 or 2 nt or containing one mismatch in the first two positions of the sgRNA binding sequence were efficient alternatives to canonical sgRNAs, increasing the number of targets in the zebrafish genome by eightfold (Moreno-Mateos et al. 2015). Notably, the activities of truncated sgRNAs are similar to those of canonical sgRNAs in ex vivo systems as well (Fu et al. 2014), supporting the use of shorter sgRNAs for genome editing in vivo.

CONCLUDING REMARKS

The CRISPR–Cas9 system has revolutionized gene targeting and genome engineering. However, using it at its full potential requires optimizations and instructions on how to apply it. In this introduction, we have reviewed two improvements that allow researchers to select the most active and convenient sgRNAs: optimization of sgRNA targeting efficiency and expansion of the potential targets in the genome. Both optimizations have been integrated into our protocol for in vivo genome targeting and can be found in CRISPRscan; see Protocol 1: Optimized CRISPR–Cas9 System for Genome Editing in Zebrafish (Vejnar et al. 2016).

ACKNOWLEDGMENTS

We thank Elizabeth Fleming and Hiba Codore for technical help; all the members of the Giraldez laboratory for intellectual and technical support; and Elizabeth Fleming, Cassandra Kontur, Timothy Johnstone, and Miler Lee for manuscript editing. The Swiss National Science Foundation (grant P2GEP3_148600 to C.E.V.), Programa de Movilidad en Áreas de Investigación priorizadas por la Consejería de Igualdad, Salud y Políticas Sociales de la Junta de Andalucía (M.A.M.-M.), the Eunice

Kennedy Shriver National Institute of Child Health and Human Development-National Institutes of Health (NIH) grant K99HD071968 (D.C.), and the NIH grants R21 HD073768 (A.J.G.), R01 HD073768 (A.J.G.), and R01 GM102251 (A.J.G.) supported our work.

REFERENCES

Bassett AR, Liu JL. 2014. CRISPR/Cas9 and genome editing in *Drosophila*. *J Genet Genomics* **41:** 7–19.

Chari R, Mali P, Moosburner M, Church GM. 2015. Unraveling CRISPR–Cas9 genome engineering parameters via a library-on-library approach. *Nat Methods* **12:** 823–826.

Cong L, Ran FA, Cox D, Lin S, Barretto R, Habib N, Hsu PD, Wu X, Jiang W, Marraffini LA, et al. 2013. Multiplex genome engineering using CRISPR/Cas systems. *Science* **339:** 819–823.

Doench JG, Hartenian E, Graham DB, Tothova Z, Hegde M, Smith I, Sullender M, Ebert BL, Xavier RJ, Root DE. 2014. Rational design of highly active sgRNAs for CRISPR-Cas9-mediated gene inactivation. *Nat Biotechnol* **32:** 1262–1267.

Esvelt KM, Mali P, Braff JL, Moosburner M, Yaung SJ, Church GM. 2013. Orthogonal Cas9 proteins for RNA-guided gene regulation and editing. *Nat Methods* **10:** 1116–1121.

Farboud B, Meyer BJ. 2015. Dramatic enhancement of genome editing by CRISPR/Cas9 through improved guide RNA design. *Genetics* **199:** 959–971.

Friedland AE, Tzur YB, Esvelt KM, Colaiacovo MP, Church GM, Calarco JA. 2013. Heritable genome editing in *C. elegans* via a CRISPR–Cas9 system. *Nat Methods* **10:** 741–743.

Fu Y, Sander JD, Reyon D, Cascio VM, Joung JK. 2014. Improving CRISPR–Cas nuclease specificity using truncated guide RNAs. *Nat Biotechnol* **32:** 279–284.

Gagnon JA, Valen E, Thyme SB, Huang P, Ahkmetova L, Pauli A, Montague TG, Zimmerman S, Richter C, Schier AF. 2014. Efficient mutagenesis by Cas9 protein-mediated oligonucleotide insertion and large-scale assessment of single-guide RNAs. *PLoS One* **9:** e98186.

Hendel A, Bak RO, Clark JT, Kennedy AB, Ryan DE, Roy S, Steinfeld I, Lunstad BD, Kaiser RJ, Wilkens AB, et al. 2015. Chemically modified guide RNAs enhance CRISPR–Cas genome editing in human primary cells. *Nat Biotechnol* **33:** 985–989.

Hou Z, Zhang Y, Propson NE, Howden SE, Chu LF, Sontheimer EJ, Thomson JA. 2013. Efficient genome engineering in human pluripotent stem cells using Cas9 from *Neisseria meningitidis*. *Proc Natl Acad Sci* **110:** 15644–15649.

Hwang WY, Fu Y, Reyon D, Maeder ML, Tsai SQ, Sander JD, Peterson RT, Yeh JR, Joung JK. 2013. Efficient genome editing in zebrafish using a CRISPR–Cas system. *Nat Biotechnol* **31:** 227–229.

Jinek M, East A, Cheng A, Lin S, Ma E, Doudna J. 2013. RNA-programmed genome editing in human cells. *eLife* **2:** e00471.

Kim S, Kim D, Cho SW, Kim J, Kim JS. 2014. Highly efficient RNA-guided genome editing in human cells via delivery of purified Cas9 ribonucleoproteins. *Genome Res* **24:** 1012–1019.

Kleinstiver BP, Prew MS, Tsai SQ, Topkar VV, Nguyen NT, Zheng Z, Gonzales AP, Li Z, Peterson RT, Yeh JR, et al. 2015. Engineered CRISPR–Cas9 nucleases with altered PAM specificities. *Nature* **523:** 481–485.

Mali P, Yang L, Esvelt KM, Aach J, Guell M, DiCarlo JE, Norville JE, Church GM. 2013. RNA-guided human genome engineering via Cas9. *Science* **339:** 823–826.

Moreno-Mateos MA, Vejnar CE, Beaudoin JD, Fernandez JP, Mis EK, Khokha MK, Giraldez AJ. 2015. CRISPRscan: Designing highly efficient sgRNAs for CRISPR–Cas9 targeting in vivo. *Nat Methods* **12:** 982–988.

Ran FA, Cong L, Yan WX, Scott DA, Gootenberg JS, Kriz AJ, Zetsche B, Shalem O, Wu X, Makarova KS, et al. 2015. In vivo genome editing using *Staphylococcus aureus* Cas9. *Nature* **520:** 186–191.

Vejnar CE, Moreno-Mateos MA, Cifuentes D, Bazzini AA, Giraldez AJ. 2016. Optimized CRISPR–Cas9 system for genome editing in zebrafish. *Cold Spring Harb Protoc* doi: 10.1101/pdb.prot086850.

Wang H, Yang H, Shivalila CS, Dawlaty MM, Cheng AW, Zhang F, Jaenisch R. 2013. One-step generation of mice carrying mutations in multiple genes by CRISPR/Cas-mediated genome engineering. *Cell* **153:** 910–918.

Wang T, Wei JJ, Sabatini DM, Lander ES. 2014. Genetic screens in human cells using the CRISPR–Cas9 system. *Science* **343:** 80–84.

Wu X, Scott DA, Kriz AJ, Chiu AC, Hsu PD, Dadon DB, Cheng AW, Trevino AE, Konermann S, Chen S, et al. 2014. Genome-wide binding of the CRISPR endonuclease Cas9 in mammalian cells. *Nat Biotechnol* **32:** 670–676.

Xu H, Xiao T, Chen CH, Li W, Meyer CA, Wu Q, Wu D, Cong L, Zhang F, Liu JS, et al. 2015. Sequence determinants of improved CRISPR sgRNA design. *Genome Res* **25:** 1147–1157.

Zetsche B, Gootenberg JS, Abudayyeh OO, Slaymaker IM, Makarova KS, Essletzbichler P, Volz SE, Joung J, van der Oost J, Regev A, et al. 2015. Cpf1 is a single RNA-guided endonuclease of a class 2 CRISPR–Cas system. *Cell* **163:** 759–771.

Cite this introduction as *Cold Spring Harb Protoc*; doi:10.1101/pdb.top090894

Protocol 1

Optimized CRISPR–Cas9 System for Genome Editing in Zebrafish

Charles E. Vejnar,[1,4] Miguel A. Moreno-Mateos,[1,4] Daniel Cifuentes,[1,3] Ariel A. Bazzini,[1] and Antonio J. Giraldez[1,2,5]

[1]*Department of Genetics, Yale University School of Medicine, New Haven, Connecticut 06510;* [2]*Yale Stem Cell Center, Yale University School of Medicine, New Haven, Connecticut 06520*

This protocol describes how to generate and genotype mutants using an optimized CRISPR–Cas9 genome-editing system in zebrafish (CRISPRscan). Because single guide RNAs (sgRNAs) have variable efficiency when targeting specific loci, our protocol starts by explaining how to use the web tool CRISPRscan to design highly efficient sgRNAs. The CRISPRscan algorithm is based on the results of an integrated analysis of more than 1000 sgRNAs in zebrafish, which uncovered highly predictive factors that influence Cas9 activity. Next, we describe how to easily generate sgRNAs in vitro, which can then be injected in vivo to target specific loci. The use of highly efficient sgRNAs can lead to biallelic mutations in the injected embryos, causing lethality. We explain how targeting Cas9 to the germline increases viability by reducing somatic mutations. Finally, we combine two methods to identify F_1 heterozygous fish carrying the desired mutations: (i) Mut-Seq, a method based on high-throughput sequencing to detect F_0 founder fish; and (ii) a polymerase chain reaction–based fragment analysis method that identifies F_1 heterozygous fish characterized by Mut-Seq. In summary, this protocol includes the steps to generate and characterize mutant zebrafish lines using the CRISPR–Cas9 genome engineering system.

MATERIALS

It is essential that you consult the appropriate Material Safety Data Sheets and your institution's Environmental Health and Safety Office for proper handling of equipment and hazardous materials used in this protocol.

RECIPES: Please see the end of this protocol for recipes indicated by <R>. Additional recipes can be found online at http://cshprotocols.cshlp.org/site/recipes.

Reagents

Agarose gel (1%, 2%, and 4% in gel running buffer of choice) and corresponding gel running buffer

We use 0.4 µg/µL ethidium bromide in all agarose gels.

AmpliScribe-T7 Flash Transcription kit (Epicentre), including all buffers

AmpliTaq DNA Polymerase with all the components (Applied Biosystems)

Annealing buffer for barcoding (2×) <R>

ATP, 10 mM (New England Biolabs)

Blue water <R>

CutSmart Buffer (New England Biolabs)

[3]Present address: Department of Biochemistry, Boston University School of Medicine, Boston, Massachusetts 02118

[4]These authors contributed equally to this work.

[5]Correspondence: antonio.giraldez@yale.edu

Copyright © Cold Spring Harbor Laboratory Press; all rights reserved

Cite this protocol as *Cold Spring Harb Protoc*; doi:10.1101/pdb.prot086850

dNTP mix (10 mM per nt) (New England Biolabs)
Ethanol (70% [prepared with RNase/DNase-free water], 95%–100% [RNA grade])
Formamide DI deionized (ultra pure) (American Bioanalytical)
mMessage mMachine SP6 Transcription Kit (Ambion) including all buffers
mMessage mMachine T3 Transcription Kit (Ambion) including all buffers
MS-222 (ethyl 3-aminobenzoate methanesulfonate salt [Sigma-Aldrich]), 500 mg/L in water
NaOH (100 mM)
NEBuffer 3.1 (New England Biolabs)
NotI (New England Biolabs)
Plasmids:
pT3TSnCas9n (Addgene 46757) (Jao et al. 2013)
pCS2-nCas9n-nanos 3′UTR (Addgene 62542) (Moreno-Mateos et al. 2015)
PCR primers (fragment analysis PCR [Step 41], Mut-Seq [Step 52], 10 µM)
Primers:
sgRNA primers:
Specific primer:
5′-*TAATACGACTCACTATA*[GG$N_{(18)}$]GTTTTAGAGCTAGAA-3′
Universal primer:
5′-AAAAGCACCGACTCGGTGCCACTTTTTCAAGTTGATAACGGACTAGCCTTATTTT AACTTGCTATTTCTAGCTCTAAAAC-3′
Mut-Seq primers:
Adapter_A: 5′(PO_4)-GATCGGAAGAGCACACGTCT-3′
Adapter_B: 5′-ACACTCTTTCCCTACACGACGCTCTTCCGATCT-3′
Adapter_C: 5′-AGGATGATACCGACCACCGAGATCTACACTCTTTCCCTACACGA-3′

In the single guide RNA (sgRNA) primers, italics indicate the T7 promoter and underlines indicate the complementary regions of the specific and universal primers. The "N(18)" sequence of the specific primer is determined as described in Steps 1 and 2.

QIAquick Gel Extraction Kit (QIAGEN)
QIAquick PCR Purification Kit (QIAGEN)
RNase/DNase-free water (Ambion)
RNeasy Mini kit (QIAGEN) including all buffers
Sodium acetate (3 M, pH 5.2)
T4 DNA ligase (New England Biolabs)
T4 DNA ligase buffer, 10× (New England Biolabs)
T4 polynucleotide kinase (PNK) (New England Biolabs)
Tris-Cl (1 M, pH 7.4)
TruSeq Barcodes: ScriptMiner Index PCR Primers (1–12) (Epicentre)
TURBO DNase (Ambion)
XbaI (New England Biolabs)
Zebrafish of desired genotypes

Equipment

Access to DNA analysis and sequencing facilities

See Steps 49, 51, and 75.

Benchtop cooler (−20°C)
Dry bath incubator
Fragment analysis software (e.g., GeneMarker)
Freezers (−20°C, −80°C)
Gel electrophoresis apparatus
Gel imager

Cite this protocol as *Cold Spring Harb Protoc*; doi:10.1101/pdb.prot086850

GMAP read aligner (download from http://research-pub.gene.com/gmap/)
Incubators (28°C, 37°C)
Microcentrifuge
Microcentrifuge tubes (1.5 mL)
Mut-Seq pipeline (download from http://protocol.crisprscan.org)

Download on a machine with Python 3.2 (or more recent). A convenient Python installer such as Anaconda (http://continuum.io/downloads) can be used.

Pasteur pipettes
PCR tubes (0.2 mL), strip or plate
sgRNA prediction tool (at http://www.crisprscan.org/)
Spectrophotometer (e.g., NanoDrop)
Thermocycler
Zebrafish embryo microinjection system

METHOD

Design of sgRNA (Fig. 1)

1. Predict sgRNA(s).

Three methods to design sgRNAs are described below; each is adapted to a specific situation. Sequences including the sgRNA primer that contains T7 promoter, sgRNA, and tail are provided (Fig. 2A).

Although the CRISPRscan model (Moreno-Mateos et al. 2015) has been designed in zebrafish and tested on both zebrafish and Xenopus, *predictions for multiple species are provided on the CRISPRscan website. These predictions directly apply the zebrafish model without further validation.*

Case 1: Targeting the coding sequence of a gene in a species included on CRISPRscan.org (Fig. 1A)

i. In CRISPRscan.org, go to the (1) "By gene" tab.

ii. Choose species of interest (2) and "Gene" in (3).

Alternatively, CRISPRscan.org can be used to target a specific transcript/isoform by selecting "Transcript" instead of "Gene" in (3).

iii. Enter gene name or ID in (4). In case this name is incomplete, choose a gene from the list of choices offered. Get sgRNAs by clicking on (5).

Results in panel (6) for the selected gene will include sgRNAs for all transcript isoforms (coordinates are reported in relation to the cDNA coding sequence). Sorting by column in panel (6) is available by clicking on column header. Presence of off-targets in the genome and canonical (GGN(18)) sites are indicated with a tick (check mark).

Case 2: Targeting an intergenic or noncoding (e.g., 3′-UTR) region in a species included on CRISPRscan.org

i. In CRISPRscan.org, go to the "UCSC tracks" tab (Fig. 1A (1)).

ii. Click on the species of interest to load tracks into the UCSC genome browser.

iii. Within the UCSC genome browser, navigate to the region of interest.

iv. Retrieve an sgRNA sequence by clicking on its name.

Case 3: Targeting a species not included on CRISPRscan.org (Fig. 1B)

i. On CRISPRscan.org, go to the (1) "Submit sequence" tab.

ii. Copy and paste your sequence of interest in (2).

iii. Click on (3) "Get sgRNAs" to sgRNAs in panel (4).

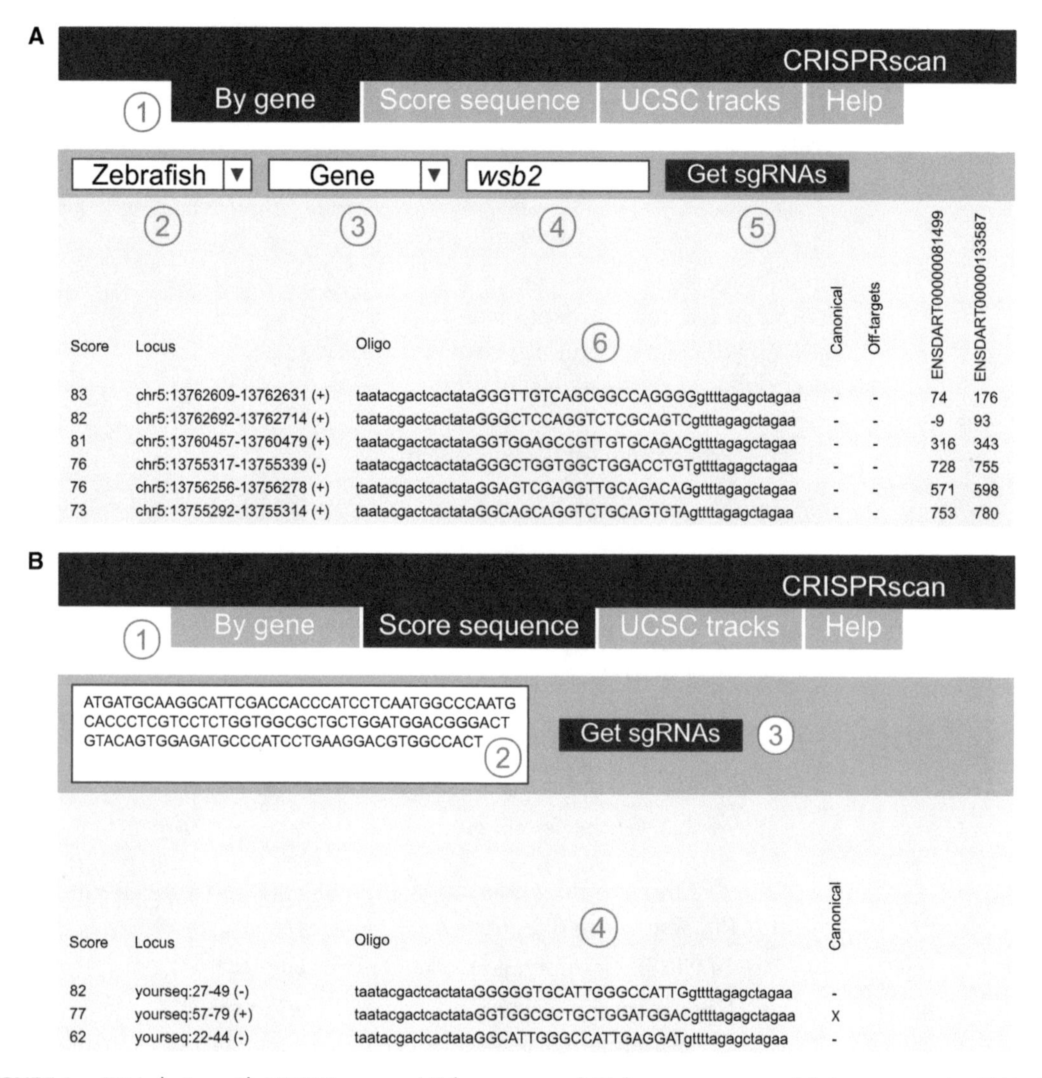

FIGURE 1. sgRNA design with CRISPRscan.org (*A*) for a gene and (*B*) for any sequence. (*A*) Precomputed sgRNAs for multiple species (1) and genes or transcripts (2) are displayed on panel (6). (*B*) User can enter a sequence (or multiple sequences in FASTA format) in (2) to identify and score potential sgRNAs.

2. Select sgRNA(s).

   Most appropriate sgRNAs should have (i) a maximal CRISPRscan score of at least 55, ideally >70 (Moreno-Mateos et al. 2015), (ii) close proximity to the targeted site, and (iii) absence of off-target sites. If there are no sgRNAs that satisfy all of these criteria (e.g., when targeting short elements such as miRNAs or transcription factor binding sites), first use a lower CRISPRscan score threshold to select an sgRNA. If no such sgRNAs exist, choose two sgRNAs with the highest possible score flanking the short element to induce a deletion that spans the two sgRNAs.

sgRNA Generation (Fig. 2A)

sgRNA DNA Template Synthesis: Fill-in PCR and DNA Purification

3. Assemble the following polymerase chain reaction (PCR) master mix on ice in a PCR tube (volumes shown are for one reaction):

   2.5 µL PCR Buffer II (10×) (AmpliTaq DNA Polymerase kit)
   0.5 µL dNTP mix (10 mM per nt)
   1.5 µL $MgCl_2$ (25 mM) (AmpliTaq DNA Polymerase kit)
   1 µL Universal primer (10 µM)

Cite this protocol as *Cold Spring Harb Protoc*; doi:10.1101/pdb.prot086850

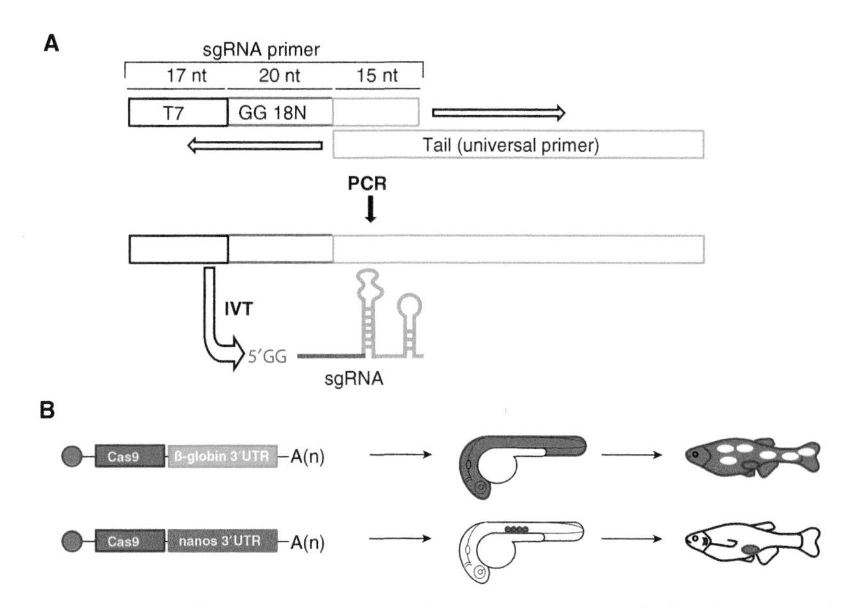

FIGURE 2. sgRNA generation and *Cas9-nanos* mutagenesis strategy. (*A*) PCR method to obtain a 117-bp template for sgRNA in vitro transcription (IVT). Each sgRNA primer contains the T7 promoter (green box), the 20 nt of the specific sgRNA–DNA binding sequence (red box), and a constant 15-nt tail which is used to anneal with an 80-nt reverse oligonucleotide (universal primer) to add the sgRNA constant 3′ end (light blue box). (*B*) Schema illustrating the *Cas9-nanos* 3′-UTR approach. Injection of *Cas9-nanos* will concentrate the expression in the germ cells (blue circles) and injection of *Cas9–β-globin* will induce mutations in the soma and generate mosaic fish. (Adapted by permission from Macmillan Publishers Ltd: *Nature Methods* [Moreno-Mateos et al. 2015], © 2015.)

1 µL sgRNA primer from Steps 1 to 2 (10 µM)
0.2 µlLAmpliTaq polymerase (5 U/µL)
18.3 µL RNase/DNase-free water
25 µL total

If two or three sgRNAs are used to target one particular gene in the same injection, the sgRNA primers can be mixed in equimolar concentrations and 1 µL of this mix (10 µM) should be used in the reaction.

4. Enter the following program in the thermocycler:

1 cycle	95°C	3 min
30 cycles	95°C	30 sec
	45°C	30 sec
	72°C	30 sec
1 cycle	72°C	5 min

5. Place the samples into the thermocycler and run the program.
6. After the program has completed, run an aliquot (5 µL of the sample) on an ethidium-bromide-stained 2% agarose gel. Visualize the PCR product (117 bp) on a standard gel imager to ensure proper amplification.
7. Clean up the rest of the PCR product (20 µL) with the QIAquick PCR Purification Kit.

   The following steps are based on the protocol for this kit, and the referenced buffers are contained in the kit.

Bind DNA to column

i. Add 5 volumes buffer PB to the PCR reaction (1 volume) and mix.

   If the color of the mixture is orange or violet, add 10 µL 3 M sodium acetate, pH 5.2, and mix. The color of the mixture will turn yellow.

ii. Place the QIAquick column in the provided 2-mL collection tube.

iii. To bind DNA, apply the sample/PB buffer mixture to the column.

iv. Centrifuge at 18,000g for 1 min.

v. Discard the flowthrough and place the column back into the same tube.

Wash DNA bound to column

vi. To wash, add 750 µL buffer PE to the column.

vii. Centrifuge at 18,000g for 1 min.

viii. Discard the flowthrough and place the column back into the same tube.

ix. To remove residual ethanol, centrifuge at 18,000g for 1 min.

Elute DNA from the column

x. Place the QIAquick column in a clean 1.5-mL microcentrifuge tube.

xi. To elute the DNA add 30 µL of RNase/DNase-free water to the center of the membrane.

xii. Let the column stand for 1 min and then centrifuge at 18,000g for 1 min at room temperature.

DNA can be stored at −20°C.

sgRNA In Vitro Transcription (IVT) Reaction

This protocol is based on the AmpliScribe-T7 Flash Transcription kit (Epicentre).

8. Vortex and thoroughly mix all the reagents.

   Keeping the reaction buffer at room temperature is very important because the spermidine in chilled 10× reaction buffer can precipitate the template DNA and causes the reaction to fail.

9. Setup IVT reaction at room temperature while keeping the enzyme solutions in a −20°C benchtop cooler (volumes shown are for one reaction). Incubate for 5–6 h at 37°C.

   6.3 µL sgRNA DNA template (from Step 7)
   1.8 µL ATP (100 mM; from AmpliScribe kit)
   1.8 µL CTP (100 mM; from AmpliScribe kit)
   1.8 µL GTP (100 mM; from AmpliScribe kit)
   1.8 µL UTP (100 mM; from AmpliScribe kit)
   2 µL DTT (100 mM) (AmpliScribe kit)
   2 µL AmpliScribe-T7 Flash 10× Reaction buffer
   0.5 µL RiboGuard RNase Inhibitor (AmpliScribe kit)
   2 µL AmpliScribe T7-Flash Enzyme Solution
   20 µL total

10. Add 1 µL TURBO DNase (2 U/µL).

11. Incubate for 20 min at 37°C.

    6.3 µL of sgRNA DNA template is ~120–150 ng.

    The reaction is also efficient if it is performed with half the volume of each component.

sgRNA Precipitation

12. Add 80 µL RNase/DNase-free water.

13. Add 10 µL of 3 M sodium acetate pH 5.2 and mix by vortexing.

14. Add 300 µL RNA-grade 95%–100% ethanol and mix by vortexing.

15. Incubate for 1 h at −80°C.

16. Centrifuge at 16,100g for 30 min at 4°C.

    A white pellet should be observed.

Cite this protocol as *Cold Spring Harb Protoc*; doi:10.1101/pdb.prot086850

17. Discard the supernatant carefully and add 750 µL 70% ethanol to rinse.
18. Centrifuge at 16,100*g* for 5 min at 4°C.
19. Repeat Steps 17–18.
20. Remove the supernatant carefully and dry the pellet 5 min to evaporate the ethanol.
21. Resuspend in 100 µL RNase/DNase-free water.
22. Dilute 5 µL in 20 µL RNase/DNase-free water (1/5 dilution). Make several aliquots.
23. Use 1 µL of the 1/5 dilution to quantify the sgRNA in a spectrophotometer.

    This protocol yields ~100 µg of sgRNA. Therefore, the concentration in the dilution sample (Step 22) should be ~200 ng/µL.

24. Store sgRNA aliquots at −80°C.

    We do not recommend freezing and thawing the sgRNAs more than two to three times.

Cas9 mRNA Production

Although regular Cas9 (Cas9-β-globin) will induce mutations in the soma and germ cells, Cas9-nanos 3′ UTR (Cas9-nanos) will concentrate the mutations in the germ cells (Fig. 2B; Koprunner et al. 2001; Mishima et al. 2006), reducing the possible toxicity or lethality caused by the mutations in the studied gene (Moreno-Mateos et al. 2015).

Linearized Plasmid Template Generation

25. To linearize Cas9 plasmids, set up the following reactions:

Cas9 (Cas9-β-globin)	Cas9-*nanos*
2 µg pT3TSnCas9n	2 µg pCS2-Cas9-*nanos*
3 µL CutSmart Buffer	3 µL NEBuffer 3.1
1 µL XbaI (10 U)	1 µL NotI (10 U)
× µL RNase/DNase-free water	× µL RNase/DNase-free water
30 µL total	30 µL total

26. Incubate for 2 h at 37°C.
27. Run the reaction in an ethidium-bromide-stained 1% agarose gel.
28. Visualize the linearized plasmid on a gel imager using long-wavelength UV and excise the fragment with a clean scalpel blade.
29. Place the gel slice in a labeled microcentrifuge tube.
30. Purify the linearized plasmid using a QIAquick Gel Extraction Kit.

    The following steps are based on the protocol for this kit, and the referenced buffers are contained in the kit.

Bind DNA to column

i. Excise the DNA fragment from thc agarose gel.
ii. Weigh the gel slice and add 3 volumes QG buffer to 1 volume gel (1 mg = 1 µL).
iii. Incubate at 50°C, vortexing every 2 min until the gel slice has dissolved completely (~6–7 min).

    If the color of the mixture is orange or violet, add 10 µL 3 M sodium acetate pH 5.2 and mix.

iv. Add 1 volume isopropanol to the sample and mix.
v. Place a QIAquick spin column in a provided 2-mL collection tube
vi. Apply the sample to the QIAquick column and centrifuge 1 min at 18,000*g* at room temperature. Discard the flowthrough and reuse the collection tube.

    Samples >750 µL should be loaded and centrifuged again.

Wash DNA bound to column

vii. Add 500 μL Buffer QG to the column.

viii. Centrifuge for 1 min at 18,000*g*. Discard the flowthrough and reuse the collection tube.

ix. Add 750 μL PE buffer to the column. Let stand 2–5 min after addition of the PE buffer.

x. Centrifuge for 1 min at 18,000*g*. Discard the flowthrough and reuse the collection tube.

xi. To remove residual PE buffer, centrifuge for 1 min at 18,000*g* at room temperature and discard the collection tube.

Elute DNA from the column

xii. Place the QIAquick column in a clean 1.5-mL microcentrifuge tube.

xiii. To elute the DNA add 30 μL of RNase/DNase-free water to the center of the membrane.

xiv. Let the column stand for 4 min and centrifuge 1 min at 18,000*g*.

IVT DNA template can be stored at −20°C.

Cas9 In Vitro Transcription and Purification

This protocol is based on the mMessage mMachine (T3 for Cas9 and SP6 for Cas9-nanos) Transcription Kit (Ambion).

31. Thoroughly vortex the 10× Reaction Buffer and the 2× NTP/CAP.

32. Store the 2× NTP/CAP (ribonucleotides) on ice, but keep the 10× Reaction Buffer at room temperature while assembling the reaction.

    Keeping the buffer at room temperature is very important because the spermidine in chilled 10× reaction buffer can precipitate the template DNA and cause the reaction to fail.

33. Setup IVT reaction at room temperature while keeping the enzyme solution in a −20°C benchtop cooler. Add the reagents in the following order (volumes shown are for one reaction):

    6 μL linearized plasmid (this is ~200–300 ng of DNA)
    10 μL NTP/CAP (2×)
    2 μL Reaction buffer (10×)
    2 μL Enzyme Mix
    20 μL total

34. Incubate for 2 h at 37°C.

35. Add 1 μL TURBO DNase (2 U/μL).

36. Incubate for 20 min at 37°C.

37. Stop the reaction by freezing the tube in liquid nitrogen and keeping it at −80°C, or continue with the mRNA purification immediately.

38. Purify the RNA using an RNeasy Mini kit (QIAGEN).

    The following steps are based on the protocol for this kit, and the referenced buffers are supplied in the kit:

Bind RNA to column

i. Add to 80 μL RNase/DNase-free water to the IVT sample.

ii. Add 350 μL Buffer RLT, mix well.

iii. Add 250 μL 96%–100% ethanol and mix by pipetting. Do not centrifuge. Proceed immediately to the next step.

iv. Transfer the sample to an RNeasy Mini spin column placed in a 2-mL collection tube. Centrifuge for 15 sec at 8000*g*. Discard the flowthrough carefully so that the column does not contact the flowthrough.

 Cite this protocol as *Cold Spring Harb Protoc*; doi:10.1101/pdb.prot086850

Wash RNA bound to column

v. Add 500 µL Buffer RPE to the column. Centrifuge for 15 sec at 8000*g*. Discard carefully the flowthrough.

vi. Add 500 µL Buffer RPE to the column. Centrifuge for 2 min at 8000*g*. Discard carefully the flowthrough.

vii. Place the RNeasy spin column in a new 2-mL collection tube. Centrifuge for 1 min at max speed to eliminate any possible carryover of buffer RPE.

Elute RNA from the column

viii. Place the RNeasy spin column in a new 1.5-mL collection tube. Add 30 µL RNase/DNase-free water to the spin column membrane.

ix. Centrifuge for 1 min at 8000*g* to elute the RNA.

x. Add the eluate from the previous step to the used column and centrifuge 1 min at 8000*g* to elute the RNA.

This step will elute more RNA and will concentrate it.

xi. Make aliquots of 5 µL and use 1 µL to quantify the mRNA concentration in a spectrophotometer.

This protocol yields ~20–30 µg of sgRNA. Therefore, the concentration should be ~500–1000 ng/µL.

xii. Store sgRNA aliquots at −80°C.

It is not recommended to thaw and freeze mRNA more than two to three times.

Injection

Cas9–nanos will concentrate the mutations in the germ cells, whereas regular Cas9 will induce mutations in the somatic and germ cells (Fig. 2B).

39. Mix sgRNA plus Cas9 mRNA in 5 µL as final volume.

Working concentration should be 100 ng/µL Cas9 mRNA and 20 ng/µL per sgRNA (100 and 20 pg of Cas9 and sgRNA(s) will be injected, respectively).

40. Inject 1 nL directly into the cell during the early one-cell stage.

As a positive control, an sgRNA (5′-GGGGAAGGTTGATTATGCAC-3′) targeting the albino *(*slc45a2*) gene (involved in pigmentation) can be used (working concentration 10 ng/µL). Using Cas9 results in 60%–70% embryos with albino-like phenotype (lack of pigmentation). Using Cas9–nanos, 70%–100% of the injected embryos show albino mutant clones in the retina.*

Several sgRNAs targeting the same locus or different loci in the same injection can be used. When more than three to four sgRNAs are used together, 10–15 ng/µL per sgRNA used is sufficient. In addition, the concentration of Cas9 can be scaled up to 200–300 ng/µL.

Injections can also be performed using the purified protein (Gagnon et al. 2014). We used 300 pg of purified protein (PNA Bio) and 200 pg sgRNA. In our hands, there was no significant difference compared with mRNA/sgRNA injections, most likely because Cas9 mRNA is already polyadenylated and is therefore translated immediately upon injection.

Analysis of CRISPR–Cas9 Mutations: (A) Using Fragment Analysis

The low cost and high efficiency of the CRISPR–Cas9 system in zebrafish allows for the generation of multiple mutant lines in parallel. Below, we describe two methods to identify mutants: (A) Fragment analysis, a fast method recommended for identifying mutant alleles with nucleotide resolution for a few targeted genes/fish (Steps 41–51) and (B) Mut-Seq, a high-throughput method designed to identify the mutant allele sequences for multiple targeted genes (Steps 52–80).

Fragment analysis is a simplified version of Sanger DNA sequencing. In the sequencing method, hundreds of fluorescently labeled PCR fragments are separated with nucleotide resolution by capillary electrophoresis and the final result is a chromatogram with an equivalent number of peaks. Each peak represents one PCR fragment and the position of the peak along the axis is proportional to its nucleotide length. In the case of fragment analysis, after capillary electrophoresis only one peak (rather than hundreds of peaks) is observed in case of wild-type samples, and any extra peak indicates the presence of mutant alleles.

Fragment analysis can be performed in (i) F_0 injected embryos with CRISPR–Cas9 to check mutagenic activity, (ii) F_1 embryos from an F_0 fish injected with CRISPR–Cas9 to identify the adult F_0 founders transmitting the mutation, or (iii) fins clipped from adult F_1 to identify the mutant allele.

Primer Design

41. Design primers to amplify an ~200–400-bp fragment covering the target(s) sgRNA sites.

    Primers should prime at least 50 nucleotides away from the target(s) sgRNA sites. One of the primers needs to be ordered with a fluorescent label at the 5′ end (usually FAM or HEX, which can be combined for multiplexing).

Embryo/Fin Collection

As a control, it is highly recommended to include at least one wild-type sample to identify the wild-type allele.

42. Collect the desired tissue using one of the following methods.

    To collect embryos

    i. Collect embryos in a PCR tube strip or plate, one embryo per tube.
    ii. Remove all water with a glass Pasteur pipette.

    To collect fins

    i. Anesthetize adult fish in diluted MS-222 (500 mg/L in water).
    ii. As soon as the fish fall unconscious, cut a small piece of the caudal fin and transfer the fin fragment with forceps to a PCR tube strip or plate.
    iii. Immediately put the fish back in a labeled, individual tank.

DNA Extraction

43. Add 100 µL of 100 mM NaOH.
44. Incubate the embryos/fin fragments for 15 min at 95°C in a dry bath incubator.
45. Immediately place the embryos on ice, neutralize with 30 µL of 1 M TrisCl, pH 7.4, and vortex to homogenize the sample.
46. In a new tube/strip, add 1 µL of the extracted DNA solution and 19 µL of water.

PCR

47. Assemble PCR reactions for each DNA sample:

    2 µL PCR Buffer II (10×) (AmpliTaq DNA Polymerase kit)
    1 µL dNTPs (10 mM)
    1.2 µL $MgCl_2$ (25 mM) (AmpliTaq DNA Polymerase kit)
    0.5 µL AmpliTaq DNA polymerase
    1 µL forward and reverse primers specific for each gene (one primer labeled with either FAM or HEX, 10 µM mix)
    1 µL genomic DNA (the 1/20 dilution prepared in Step 46)
    13.3 µL RNase/DNase-free water
    20 µL total

48. Perform PCR using the following program:

1 cycle	95°C	2 min
34 cycles	95°C	10 sec
	50°C	30 sec
	72°C	30 sec
1 cycle	72°C	5 min

49. Assemble a PCR (0.2-mL) tube with 1 µL PCR product plus 9 µL formamide DI deionized and submit it to a DNA analysis facility for fragment analysis.

    One FAM- and one HEX-labeled fragment can be combined in a single tube to reduce costs.

Cite this protocol as *Cold Spring Harb Protoc*; doi:10.1101/pdb.prot086850

50. Analyze the data using GeneMarker (SoftGenetics) or other fragment analysis software. Compare the size of the wild-type copy to the targeted ones and identify the samples showing DNA deletion or insertion.

    A DNA insertion will produce peaks of a larger size than wild type, while a DNA deletion will produce peaks of a smaller size compared with the wild-type peak (Fig. 3).

51. Sequence the PCR fragments from the desired samples to obtain the sequence of the allele.

Analysis of CRISPR–Cas9 Mutations: (B) Using Mut-Seq

Primer Design

52. Design primers to amplify a 130- to 150-bp fragment surrounding each sgRNA site.

    Primers should prime at least 30–50 nucleotides away from the target(s) sgRNA sites. Multiple target sites can be analyzed in the same PCR fragment if they are <150 nucleotides apart.

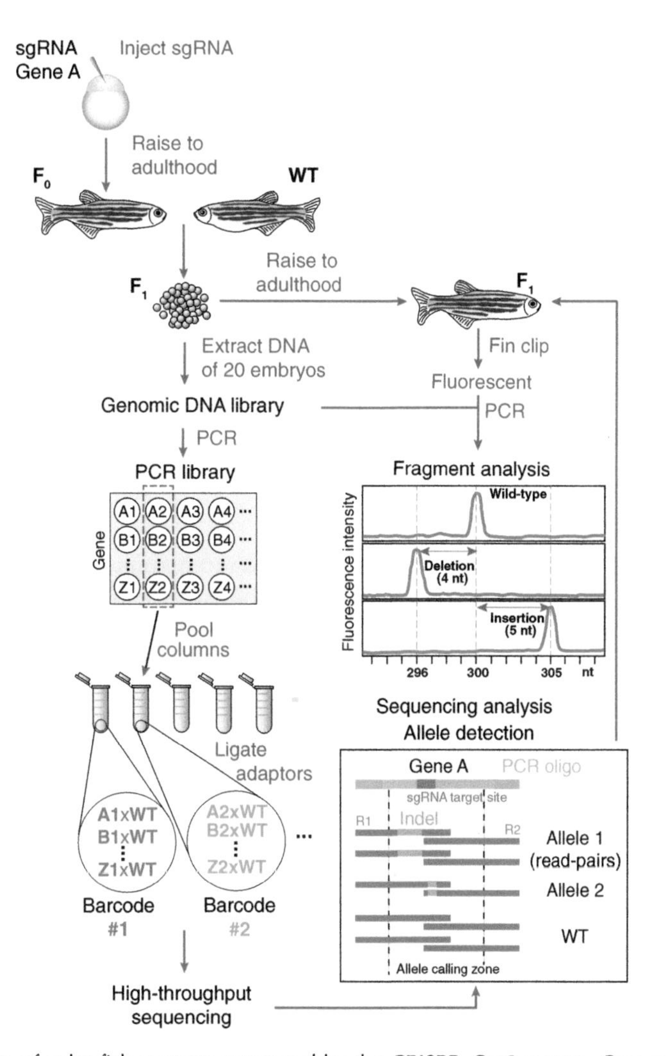

FIGURE 3. Identification of zebrafish mutants generated by the CRISPR–Cas9 system. One-cell-stage embryos are injected with the sgRNA targeting the desired gene (Gene A in the example). Those embryos are raised to adulthood (F_0) and outcrossed to wild-type (WT) fish. Twenty F_1 embryos are collected for DNA extraction and genotyping, while the remaining embryos are raised to adulthood. PCR amplicons covering the sgRNA target sites are generated for each gene target. Amplicons coming from different genes are pooled, barcoded, and sequenced, and then mutations are identified using Mut-Seq. Fragment analysis on fin clips is used to identify the F_1 heterozygous fish carrying alleles revealed by Mut-Seq.

Embryo Production and Collection

Day 0

53. In the afternoon, set crosses of F_0 CRISPR–Cas9-injected fish targeting (as example, gene A) to wild-type fish in crossing tank without using dividers (Fig. 3).

    To reduce the number of crosses, F_0 CRISPR-injected fish targeting one locus (gene A) can be crossed to other F_0 CRISPR-injected fish targeting another locus (Gene B). At the end of the pipeline, it will uniquely identify the founder fish for each gene. However, once the F_0 is identified, it will be outcrossed to wild-type to clean the line of other mutations.

Day 1

54. Collect embryos ~2 h after the fish facility lights are turned on.
55. Place the embryos in a Petri dish with blue water and label it with the same identifier as the breeding tank holding the parents.
56. Place the plate with the embryos in an incubator at 28°C.
57. Set aside the breeding tank with the parents in a safe place in the fish facility.
58. Approximately 6 h after fertilization (at shield stage), collect 20 embryos from each clutch in a separate 1.5-mL tube (Fig. 3).

    Grow the rest of the F_1 embryos to adulthood, because if the parent is carrying a germline mutation, these F_1 may carry the allele and can be genotyped as adults by fin clip and fragment analysis, to identify heterozygous fish and establish the mutant line.

DNA Extraction and PCR

59. Follow the steps explained above for DNA extraction and PCR (Steps 43–48).

    A FAM/HEX-labeled primer is not required. It is important to use a DNA polymerase that introduces A-overhangs, because they will be used for ligating the adaptors. Alternatively, the PCR can be done with a proofreading polymerase and include an extra step of incubation with Taq *polymerase to introduce the A-overhangs.*

Sample Pooling

60. To reduce the number of samples to manipulate, pool the PCRs from clutches of different gene crosses to be sequenced using the same sequencing barcode.

    As each PCR product will be specific for one specific genomic locus, it will be straightforward to identify the pair of origin even if they share the same sequencing barcode (Fig. 3).

    Fish targeted in Gene A can be pooled with fish targeted in Gene B, C, ... Z, but other fish targeted in the same gene should never be pooled to avoid misidentification of the parent fish of origin (it will be impossible to assign the identified mutations in Gene A to the correct fish founder).

Phosphorylation

Perform Steps 61–65 on each PCR pool.

61. Purify the PCR using the QIAquick PCR Purification Kit as described above (Step 7.i–7.x).
62. Elute the PCR product as in Steps 7.xi–7.xii, but now use 45 µL of water.
63. Add 5 µL of 10× T4 DNA ligase buffer and 2 µL of T4 PNK.
64. Incubate for 30 min at 37°C.
65. Heat-inactivate for 20 min at 60°C.

Barcoding

Each phosphorylated PCR pool (Steps 61–65) will be barcoded with Illumina barcodes.

 Cite this protocol as *Cold Spring Harb Protoc*; doi:10.1101/pdb.prot086850

Preparation of adapter

66. Mix:

    12.5 µL of Adapter_A (100 µM)
    12.5 µL of Adapter_B (100 µM)
    25 µL of 2× annealing buffer for barcoding

67. Heat for 1 min at 95°C and add to a floater in a beaker of boiling water. Let the water cool down to room temperature.

    The final concentration of the mixture is 25 µM.

Ligation of both adaptors to the PCR pool

68. After Step 67, add the following to the PCR pool from Steps 61–65:

    0.5 µL of 10 mM ATP
    2 µL adapter duplex (25 µM stock from previous step)
    2 µL of T4 DNA ligase

69. Incubate 2 h at room temperature or overnight at 16°C.
70. Run a 2% agarose gel and cut out the ligated band (your PCR product +55 bp).
71. Slice the band from the gel and extract the DNA using the gel extraction kit (Step 30).

Final PCR to add barcodes

72. Perform PCR as described before (Steps 47–48) using Adapter_C and TruSeq barcodes. Use 2 µL of the ligated PCR pools from the previous step (Step 71).

    Setup three PCR reactions per sample at different cycles (start by testing 8, 10, and 12 cycles) to select the number of cycles that do not overamplify the sample.

73. Run a 2% agarose gel and cut out the desired band (your PCR product +128 bp)
74. Extract the DNA using the gel extraction protocol (Step 30).
75. Sequence in an Illumina Hi-Seq or Mi-Seq sequencer, paired end, with at least 75-nucleotide reads.

Sequence analysis

76. Install the Mut-Seq analysis pipeline:

    i. Download the mutseq program (http://protocol.crisprscan.org) on a machine with Python 3.2 (or more recent).

       A convenient Python installer such as Anaconda (http://continuum.io/downloads) can be used.

    ii. Download the GMAP read aligner (http://research-pub.gene.com/gmap/) and install it.

77. Prepare Mut-Seq input:

    i. Organize sequencing FASTQ files by separating the FASTQ files for each sample (containing pooled PCRs) into folders (e.g., Pool1, Pool2).
    ii. Create a CSV (comma-separated value) file describing genes, oligo sequences to amplify them and in which sequencing pool they were added.

       All sequences must be 5′ to 3′. The first line of the final CSV must be a header line.

```
20.1% 350210 pair(s)
  TTCAAGTCTAAATCTTTTTCTAGATTTAATTAGCTCAGAAAAATATATATGACTTGTTTAATTTAGTATGATCATTACATAAATTATTTATTTGTAAGTGC
  TTCAAGTCTAAATCTTTTTCTAGATTTAATTAGCTCAGAAAAATAT----GACTTGTTTAATTTAGTATGATCATTACATAAATTATTTATTTGTAAGTGC

6.2% 102102 pair(s)
  TTCAAGTCTAAATCTTTTTCTAGATTTAATTAGCTCAGAAAAATATATATGACTTGTTTAATTTAGTATGATCATTACATAAATTATTTATTTGTAAGTGC
  TTCAAGTCTAAATCTTTTTCTAGATTTAATTAGCTCAGAAAAATATATA---CTTGTTTAATTTAGTATGATCATTACATAAATTATTTATTTGTAAGTGC
```

FIGURE 4. Mut-Seq program output for alleles of 4-nt deletion (top) and 3-nt deletion (bottom). In this example, the second allele is a 3-nt deletion, which results in an in-frame mutation inside a coding sequence. In this case, although highly frequent, fish bearing this allele have to be discarded. If targeting a noncoding region, fish bearing such an allele could be kept.

Example data:

Locus name	Pool (sequencing folder)	Forward oligo sequence	Reverse oligo sequence	Amplicon sequence
GeneA	Pool1,Pool2,Pool3	ACCA...	TGAC...	ACCA...GTCA
GeneB	Pool2,Pool3	TCCG...	ATGG...	TCCG...TACC

Corresponding CSV file:

"GeneA","Pool1,Pool2,Pool3","ACCA...","TGAC...","ACCA...GTCA"

"GeneB","Pool2,Pool3","TCCG...","ATGG...","TCCG...TACC"

78. Start the mutseq program. Specify the path to the CSV file created at the previous step using the –*path_loci* option and the path to the sequencing FASTQ files using the –*path_data* option.

The report is saved to the file specified using the –report option. For example:

Executing the mustseq program depends on installation. Please refer to http://protocol.crisprscan.org for details./mutseq.py-path_loci loci.csv-path_data /data/seq/flowcell1/Unaligned/Project_xx-report report.txt

All indels within the "Allele calling zone" (Fig. 3, lower right) are collapsed and counted with wild-type read-pairs. Allele frequency is estimated by the ratio of the number of collapsed read-pairs with same indels to the number of wild-type read-pairs.

79. Analyze the output of the mutseq program to select appropriate mutations.

For each locus, it will return the following:

- *Number of mutant, wild-type, and aligned (i.e., aligned to the locus but neither mutant nor wild type, for example PCR dimers) read-pairs.*
- *Alleles sorted by frequency (see Fig. 4).*

80. Proceed to fragment analysis to identify fish bearing the allele identified with Mut-Seq.

If the mutation is an insertion or deletion of over 10 nucleotides, the downstream analysis of the mutant line can be streamlined analyzing the genotyping PCRs by 4% agarose gel electrophoresis. This procedure will save the time and costs of fragment analysis. The only requirement is to design primers to generate PCR amplicons between 120 and 150 nucleotides long and perform the PCR reaction as described in Steps 47–48 (no need for FAM or HEX primer). In these conditions, a standard 4% agarose gel electrophoresis will have enough resolution to separate the mutant and wild-type bands.

RECIPES

Annealing Buffer for Barcoding (2×)

20 mM Tris-Cl, pH 7.8
100 mM NaCl
0.2 mM EDTA

Store for up to 1 yr at room temperature.

Blue Water

Methylene blue, 20 mL (0.1% [w/v] stock in H_2O)
Instant Ocean sea salt (Instant Ocean), 6 g
H_2O, 20 L

 Cite this protocol as *Cold Spring Harb Protoc*; doi:10.1101/pdb.prot086850

ACKNOWLEDGMENTS

We thank Elizabeth Fleming and Hiba Codore for technical help, and all the members of the Giraldez laboratory for intellectual and technical support. We also thank Elizabeth Fleming, Cassandra Kontur, Timothy Johnstone, and Miler Lee for manuscript editing. The Swiss National Science Foundation (grant P2GEP3_148600 to C.E.V), Programa de Movilidad en Áreas de Investigación priorizadas por la Consejería de Igualdad, Salud y Políticas Sociales de la Junta de Andalucía (M.A.M.-M.), the Eunice Kennedy Shriver National Institute of Child Health and Human Development–National Institutes of Health (NIH) grant K99HD071968 (D.C.), and NIH grants R21 HD073768 (A.J.G.), R01 HD073768 (A.J.G.), and R01 GM102251 (A.J.G.) supported this work.

REFERENCES

Gagnon JA, Valen E, Thyme SB, Huang P, Ahkmetova L, Pauli A, Montague TG, Zimmerman S, Richter C, Schier AF. 2014. Efficient mutagenesis by Cas9 protein-mediated oligonucleotide insertion and large-scale assessment of single-guide RNAs. *PLoS One* **9:** e98186.

Jao LE, Wente SR, Chen W. 2013. Efficient multiplex biallelic zebrafish genome editing using a CRISPR nuclease system. *Proc Natl Acad Sci* **110:** 13904–13909.

Koprunner M, Thisse C, Thisse B, Raz E. 2001. A zebrafish *nanos*-related gene is essential for the development of primordial germ cells. *Genes Dev* **15:** 2877–2885.

Mishima Y, Giraldez AJ, Takeda Y, Fujiwara T, Sakamoto H, Schier AF, Inoue K. 2006. Differential regulation of germline mRNAs in soma and germ cells by zebrafish miR-430. *Curr Biol* **16:** 2135–2142.

Moreno-Mateos MA, Vejnar CE, Beaudoin JD, Fernandez JP, Mis EK, Khokha MK, Giraldez AJ. 2015. CRISPRscan: Designing highly efficient sgRNAs for CRISPR–Cas9 targeting in vivo. *Nat Methods* **12:** 982–988.

CHAPTER 10

Editing the Mouse Genome Using the CRISPR–Cas9 System

Adam Williams,[1,6,7] Jorge Henao-Mejia,[2,3,6,7] and Richard A. Flavell[4,5,7]

[1]*The Jackson Laboratory for Genomic Medicine, Department of Genetics and Genome Sciences, University of Connecticut Health Center, Farmington, Connecticut 06032;* [2]*Institute for Immunology, Perelman School of Medicine, University of Pennsylvania, Philadelphia, Pennsylvania 19104;* [3]*Division of Transplant Immunology, Department of Pathology and Laboratory Medicine, Children's Hospital of Philadelphia, University of Pennsylvania, Philadelphia, Pennsylvania 19104;* [4]*Department of Immunobiology, Yale University School of Medicine, New Haven, Connecticut 06520;* [5]*Howard Hughes Medical Institute, Yale University School of Medicine, New Haven, Connecticut 06520*

The ability to modify the murine genome is perhaps one of the most important developments in modern biology. However, traditional methods of genomic engineering are costly and relatively clumsy in their approach. The use of programmable nucleases such as zinc finger nucleases and transcription activator-like effector nucleases significantly improved the precision of genome-editing technology, but the design and use of these nucleases remains cumbersome and prohibitively expensive. The CRISPR–Cas9 system is the next installment in the line of programmable nucleases; it provides highly efficient and precise genome-editing capabilities using reagents that are simple to design and inexpensive to generate. Furthermore, with the CRISPR–Cas9 system, it is possible to move from a hypothesis to an in vivo mouse model in less than a month. The simplicity, cost effectiveness, and speed of the CRISPR–Cas9 system allows researchers to tackle questions that otherwise would not be technically or financially viable. In this introduction, we discuss practical considerations for the use of Cas9 in genome engineering in mice.

Cas9 BACKGROUND AND PRINCIPAL COMPONENTS

Genetically modified mice are a cornerstone of biomedical research as they provide essential tools to understand gene function and to model complex human diseases. Until recently, genetically engineered mice were generated through genetic modification of mouse embryonic stem (ES) cells by homologous recombination. Targeted ES cells are expanded and injected into wild-type mouse blastocysts with the expectation that they will contribute to the germline of chimeric mice. Chimeric mice are then bred to wild-type mice to generate progeny containing the targeted locus (Thomas and Capecchi 1987). This process is extremely costly, time-consuming and, in some cases, uncertain. Although, this procedure usually takes 9–12 mo, the generation of mice carrying multiple targeted loci or challenging targeting locations can substantially add more time, effort, and economic cost.

In the last decade, different methods have been developed to generate mutant mice in a rapid and efficient manner. The most successful approaches use programmable nucleases such as zinc finger nucleases (ZFNs) and transcription activator-like effector nucleases (TALENs) (Boch et al. 2009) injected directly into mouse one-cell embryos, a procedure that greatly accelerates the process of

[6]These authors contributed equally to this work.

[7]Correspondence: adam.williams@jax.org, jhena@mail.med.upenn.edu, richard.flavell@yale.edu

Copyright © Cold Spring Harbor Laboratory Press; all rights reserved

Cite this introduction as *Cold Spring Harb Protoc*; doi:10.1101/pdb.top087536

generating genetically modified mice by avoiding the use of ES cells (for a recent review, see Kim and Kim 2014). Once injected, these nucleases have the capability to generate double-strand breaks (DSBs) at predefined sites in the genome that are then repaired by error-prone nonhomologous end joining (NHEJ), resulting in either insertion or deletion (indel) mutations; indels located within protein-coding exons can cause frameshifts resulting in a knockout allele (Kim and Kim 2014). Alternatively, if a single-stranded DNA (ssDNA) or a circular donor plasmid with homology regions flanking, the DSB is introduced into the one-cell embryo in combination with these nucleases, a defined DNA sequence can be inserted into the genome by high-fidelity homology-directed repair (HDR), allowing the generation of knock-in mice carrying point mutations, tags, conditional alleles, or fluorescent proteins (Kim and Kim 2014).

The most recently developed genome-editing tool is the CRISPR-associated protein 9 (Cas9) nuclease. The CRISPR–Cas system functions as an RNA-based adaptive immune system in bacteria and archaea (Barrangou et al. 2007). In *Streptococcus pyogenes*, a type II CRISPR–Cas system composed of Cas9, CRISPR RNAs (crRNAs), and a *trans*-activating crRNA (tracrRNA) target and degrade nucleic acids from foreign plasmids or bacteriophages (Deltcheva et al. 2011; Jinek et al. 2012). In this system, the Cas9 nuclease is guided to invading foreign nucleic acids by crRNAs that are partially complementary to the target sequence, and the transcRNA plays a pivotal structural role for the proper activity of the Cas9 nuclease. The repurposing of Cas9 to generate site-specific DSBs in mammalian genomes was a turning point in genome editing (Cho et al. 2013; Cong et al. 2013; Mali et al. 2013b; Wang et al. 2013; Yang et al. 2013). Part of this repurposing was the fusion of the crRNA and tracrRNA to form a single-guide RNA (sgRNA) (Jinek et al. 2012). To direct Cas9 to a specific genomic region, the sgRNA is designed so that the 20 nucleotides at its 5′ end are homologous to the genomic target sequence. In addition, the genomic sequence must be immediately followed by a protospacer adjacent motif (PAM) sequence, a 3-bp (NGG) motif present in the target sequence but not the sgRNA (Fig. 1).

Like previous programmable nucleases, CRISPR–Cas9 provides highly efficient and precise genome editing capabilities in mice (Wang et al. 2013; Yang et al. 2013). However, the significant advantage that Cas9 offers is that it uses reagents that are simple, inexpensive, and quick to design and generate. Furthermore, it is also significantly more efficient than ZFNs and TALENs (Yasue et al. 2014). In addition, the CRISPR–Cas9 system allows many targeting applications, such as the targeting of multiple loci simultaneously, the generation of conditional alleles, and the production of mice carrying endogenous reporters (Wang et al. 2013; Yang et al. 2013). Moreover, these modifications

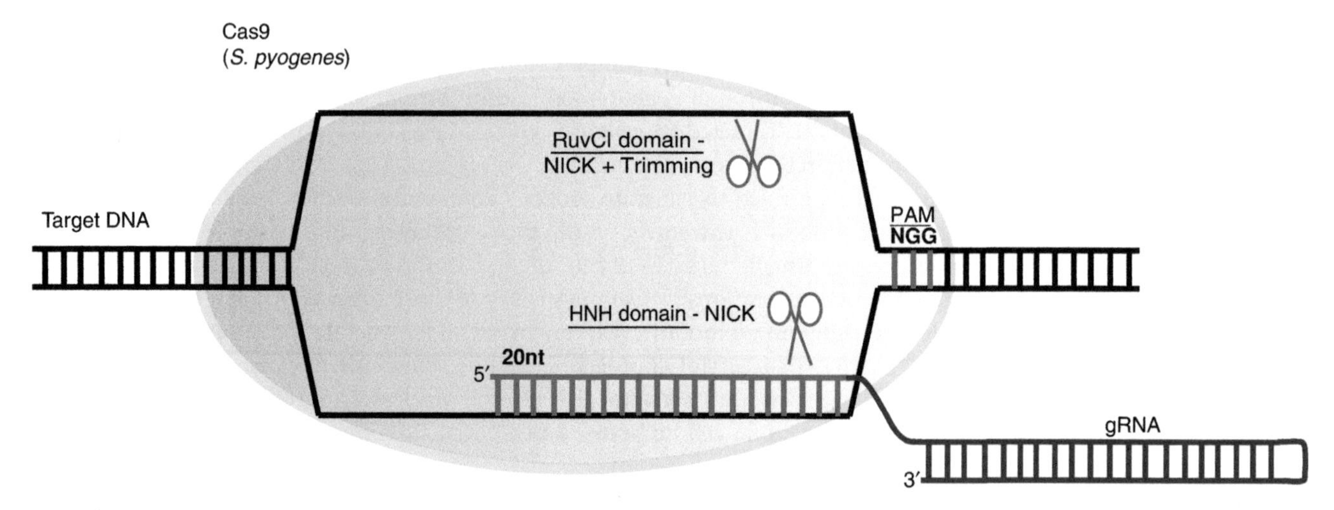

FIGURE 1. Basic components of the Cas9 system. The Cas9 nuclease generates DSBs by using its two catalytic domains (HNH and RuvCI) to cleave each strand of a DNA target site next to a PAM sequence (red) and matching the 20-nucleotide sequence of the guide RNA (gRNA). The sgRNA includes a fused RNA sequence derived from CRISPR RNA and the *trans*-activating crRNA that binds and stabilizes the Cas9 nuclease.

Cite this introduction as *Cold Spring Harb Protoc*; doi:10.1101/pdb.top087536

can be made in pure inbred strains of mice (e.g., C57BL/6) as well as directly in established mutant strains, dramatically reducing the time and cost required to generate/modify complex animal models.

Since adopting this revolutionary technology in late 2012, we have generated more than one hundred novel genetically engineered mouse strains. In concordance with previous reports, we have observed high success rates in all the potential types of genome targeting events. In Protocol 1: Generation of Genetically Modified Mice Using the CRISPR–Cas9 Genome-Editing System (Henao-Mejia et al. 2016), we describe in detail the optimal conditions to generate mice carrying point mutations, chromosomal deletions, conditional alleles, fusion tags, or endogenous reporters.

Cas9 GENOME-EDITING APPLICATIONS

In this section we will discuss general considerations for the main Cas9-mediated genome-editing applications, as outlined below (Fig. 2).

Gene Knockout through Indel Generation

One main use of genome modification has been in the generation of gene knockout animals. DSBs generated by Cas9 are repaired by the error-prone NHEJ pathway, resulting in indel generation. When targeted within the coding region of a gene indels frequently results in a frameshift and loss of function. This is the most simple and efficient form of Cas9-mediated genome editing, requiring only injection of Cas9 and a single sgRNA.

Point Mutations/Small Insertions

A powerful use of the Cas9 system is the precise editing of the mouse genome to introduce specific nucleotide changes (Wang et al. 2013). This enables disease modeling by allowing exact nucleotide changes engineered to mimic human disease mutations. In addition, the creation/destruction of specific genomic sequences, such as transcription factor binding sites, allows interrogation of their function. Cas9 can also be used to introduce short artificial sequences such as hemagglutinin tags or *loxP* sites into the genome at precise locations (Yang et al. 2013). Precise genome-editing requires three components: Cas9, an sgRNA, and an ssDNA oligo containing the desired nucleotide modifications. As described above, Cas9 is directed by an sgRNA to generate a DSB at a specific location in the genome. In the presence of an ssDNA oligo with homology flanking the DSB, the host DNA repair machinery is able to perform HDR using the donor oligo as a template; any mutation/artificial sequence included in the donor oligo will be copied into the genome at this exact location. Donor ssDNA oligos typically contain 50–60 bases either side of the region to be edited. The addition of phosphorothioate linkages at the 5′ and 3′ terminal nucleotides can increase in vivo stability of the oligos, potentially increasing the efficiency of the reaction. Using this approach we have successfully inserted an artificial 100-nt sequence into the genome.

To prevent Cas9 from recutting the edited sequence and introducing an indel, sgRNAs should be designed to place the modified region as close to the PAM sequence as possible. This increases the chance that the modification is correctly incorporated, as well as reducing the probability of Cas9 recutting (by altering the sgRNA-binding site). When making modifications at multiple sites, such as flanking an exon with *loxP* sequences, a separate sgRNA and donor oligo is required for each modification site. It is important to remember that these are independent targeting events and might not occur on the same copy of the chromosome and therefore can segregate on breeding.

Large Deletions

For some applications, the ablation of large chromosomal regions is desired. Such deletions are difficult to achieve with classical targeting methodologies, often requiring multiple rounds of recombination in ES cells (Hacisuleyman et al. 2014). However, large deletions are relatively

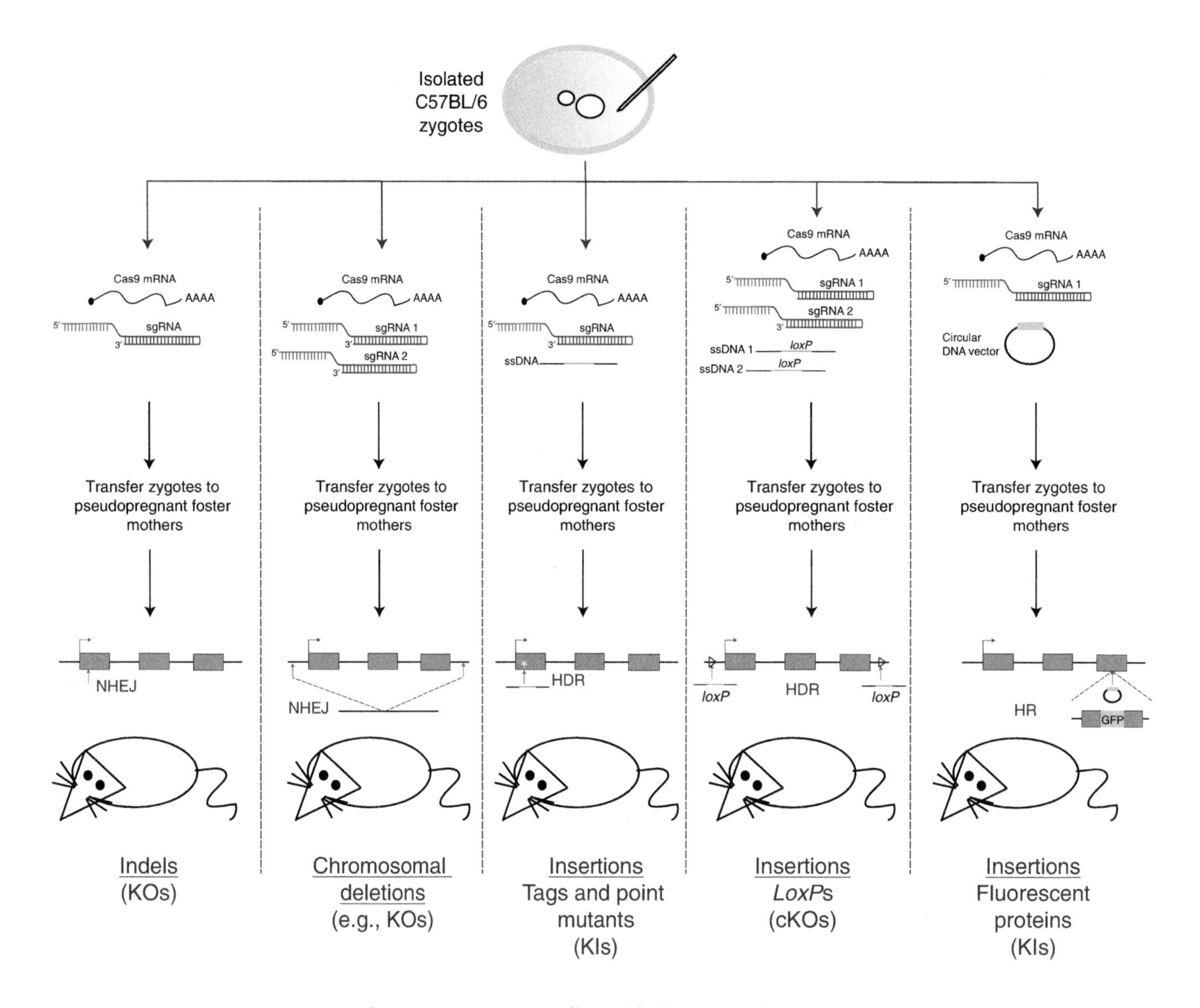

FIGURE 2. Procedures to generate genetically modified mice using the CRISPR–Cas9 genome-editing system. Isolated zygotes are co-injected with Cas9 mRNA and sgRNAs to generate mice carrying indel mutations or targeted chromosomal deletions (i.e., knockouts [KOs]). Alternatively, the Cas9 mRNA and sgRNAs are co-injected in combination with donor ssDNAs or circular plasmids to generate mice harboring point mutations, tags, *loxP* sites, or large DNA fragments such as a fluorescent protein (KIs, knock-ins; cKOs, conditional knockouts.)

simple to achieve using the CRISPR–Cas9 system. By using two sgRNAs to generate DSBs flanking a region of interest, it is possible to efficiently delete the intervening sequence through the NHEJ repair process (Yang et al. 2013; Krishnaswamy et al. 2015). Although efficiency is likely influenced by the linear distance between the sgRNAs (and potentially the three-dimensional structure of the genome), we have deleted genomic regions of up to 200-kb using this approach.

Large Insertions

To introduce large DNA sequences (e.g., fluorescent proteins) at precise locations, a plasmid that encodes the DNA sequence to be inserted flanked by >2 kb of homology is general used (Yang et al. 2013). However, we have been able to introduce large fragments of DNA with smaller homology regions (~500 bp). The DNA sequence of interest should be inserted as close as possible to the generated DSB, and if possible the sgRNA target sequence should be modified

Cite this introduction as *Cold Spring Harb Protoc*; doi:10.1101/pdb.top087536

in the targeting vector to prevent cutting of the donor DNA or recutting of the genome after HDR. We have had success using both cytoplasmic and pronuclear injections for large insertions. However, to reduce random insertion, it is best to use circular plasmids rather than linearized DNA.

DESIGN OF sgRNAs TO MAXIMIZE CUTTING AND MINIMIZE OFF-TARGETS

The design of sgRNAs is relatively simple, requiring only that the 20-nt homology be immediately followed in the genome by an NGG PAM sequence. However, certain nucleotides are favored/disfavored at different positions along the sgRNA and this should be considered during the sgRNA design (Doench et al. 2014). Most importantly, careful selection of sgRNAs is paramount to minimize potential off-target cutting; candidate sgRNAs should be blasted against the genome: an ideal sgRNA would only be present as a single site within the genome and should have at least five mismatches to any similar sites in the genome. A number of websites offer free tools to assist with the sgRNA design. The use of truncated sgRNA has been shown to reduce off-target cutting; however, we have little experience with this approach (Fu et al. 2014). Alternatively, off-target effects can be reduced by using the nickase Cas9 mutant, which can only generate DSBs when two sgRNA targets are close to each other (Mali et al. 2013a). Nickase can be used in place of regular Cas9 in any of the applications described above; however, this requires replacing every sgRNA with two sgRNAs (Lee and Lloyd 2014; Rong et al. 2014; Shen et al. 2014). We have had similar success rates using the nickase Cas9 mutant when compared with the wild-type Cas9 in the generation of mice with indel mutations or small deletions.

CONSIDERATIONS FOR SCREENING

For screening indels or small insertions, a simple PCR across the targeted region followed by a Surveyor assay can be used for initial screening. To confirm sequence of mutated alleles, cloning of the genotyping product followed by sequencing is often required. For genotyping large deletions, primers spanning the excised region provide a simple assay for deletion. However, NHEJ across large deletions can result in loss of additional nucleotides proximal to each DSB; therefore, PCR primers should be placed at least 100-bp outside of the expected cut sites. Although founder animals may sometimes carry homozygous modifications, they are frequently mosaic, with different cells in the same animal carrying different modifications; we have detected founders carrying five to six unique alleles. It is therefore important to cross founder animals to wild-type mice and then screen the various alleles after segregation. This is especially important when screening to select animals in which *loxP* sites have integrated on the same copy of the chromosome.

CONCLUSIONS AND FUTURE PERSPECTIVES

The CRISPR–Cas9 system has revolutionized genome engineering, overcoming many of the problems associated with previous programmable nucleases. However, Cas9-mediated targeting is still limited to sites containing a PAM sequence (NGG). An expanding toolbox of CRISPR–Cas from different bacterial species, with alternative PAM sequence requirements, will provide greater targeting flexibility (Hou et al. 2013). Finally, the current bottleneck in Cas9 targeting is the highly technical and time-consuming process of microinjection. In the future, high-throughput methods of delivery will replace microinjection, opening up this technology to even more labs.

ACKNOWLEDGMENTS

We would like to thank Jon Alderman for his technical assistance and discussions. We would also like to thank Caroline Lieber for her assistance in the preparation and submission of the manuscript. This work was supported in part by the National Institutes of Health National Institute of Allergy and Infectious Diseases (NIAID), grant 1R21AI110776-01 (A.W. and R.A.F).

REFERENCES

Barrangou R, Fremaux C, Deveau H, Richards M, Boyaval P, Moineau S, Romero DA, Horvath P. 2007. CRISPR provides acquired resistance against viruses in prokaryotes. *Science* **315:** 1709–1712.

Boch J, Scholze H, Schornack S, Landgraf A, Hahn S, Kay S, Lahaye T, Nickstadt A, Bonas U. 2009. Breaking the code of DNA binding specificity of TAL-type III effectors. *Science* **326:** 1509–1512.

Cho SW, Kim S, Kim JM, Kim JS. 2013. Targeted genome engineering in human cells with the Cas9 RNA-guided endonuclease. *Nat Biotechnol* **31:** 230–232.

Cong L, Ran FA, Cox D, Lin S, Barretto R, Habib N, Hsu PD, Wu X, Jiang W, Marraffini LA, et al. 2013. Multiplex genome engineering using CRISPR/Cas systems. *Science* **339:** 819–823.

Deltcheva E, Chylinski K, Sharma CM, Gonzales K, Chao Y, Pirzada ZA, Eckert MR, Vogel J, Charpentier E. 2011. CRISPR RNA maturation by trans-encoded small RNA and host factor RNase III. *Nature* **471:** 602–607.

Doench JG, Hartenian E, Graham DB, Tothova Z, Hegde M, Smith I, Sullender M, Ebert BL, Xavier RJ, Root DE. 2014. Rational design of highly active sgRNAs for CRISPR-Cas9-mediated gene inactivation. *Nat Biotechnol* **32:** 1262–1267.

Fu Y, Sander JD, Reyon D, Cascio VM, Joung JK. 2014. Improving CRISPR-Cas nuclease specificity using truncated guide RNAs. *Nat Biotechnol* **32:** 279–284.

Hacisuleyman E, Goff LA, Trapnell C, Williams A, Henao-Mejia J, Sun L, McClanahan P, Hendrickson DG, Sauvageau M, Kelley DR, et al. 2014. Topological organization of multichromosomal regions by the long intergenic noncoding RNA Firre. *Nat Struct Mol Biol* **21:** 198–206.

Henao-Mejia J, Williams A, Rongvaux A, Stein J, Hughes C, Flavell R. 2016. Generation of genetically modified mice using the CRISPR–Cas9 genome-editing system. *Cold Spring Harb Protoc* doi: 10.1101/pdb.prot090704.

Hou Z, Zhang Y, Propson NE, Howden SE, Chu LF, Sontheimer EJ, Thomson JA. 2013. Efficient genome engineering in human pluripotent stem cells using Cas9 from *Neisseria meningitidis*. *Proc Natl Acad Sci* **110:** 15644–15649.

Jinek M, Chylinski K, Fonfara I, Hauer M, Doudna JA, Charpentier E. 2012. A programmable dual-RNA-guided DNA endonuclease in adaptive bacterial immunity. *Science* **337:** 816–821.

Kim H, Kim JS. 2014. A guide to genome engineering with programmable nucleases. *Nat Rev Genet* **15:** 321–334.

Krishnaswamy JK, Singh A, Gowthaman U, Wu R, Gorrepati P, Sales Nascimento M, Gallman A, Liu D, Rhebergen AM, Calabro S, et al. 2015. Coincidental loss of DOCK8 function in NLRP10-deficient and C3H/HeJ mice results in defective dendritic cell migration. *Proc Natl Acad Sci* **112:** 3056–3061.

Lee AY, Lloyd KC. 2014. Conditional targeting of Ispd using paired Cas9 nickase and a single DNA template in mice. *FEBS Open Bio* **4:** 637–642.

Mali P, Aach J, Stranges PB, Esvelt KM, Moosburner M, Kosuri S, Yang L, Church GM. 2013a. CAS9 transcriptional activators for target specificity screening and paired nickases for cooperative genome engineering. *Nat Biotechnol* **31:** 833–838.

Mali P, Yang L, Esvelt KM, Aach J, Guell M, DiCarlo JE, Norville JE, Church GM. 2013b. RNA-guided human genome engineering via Cas9. *Science* **339:** 823–826.

Rong Z, Zhu S, Xu Y, Fu X. 2014. Homologous recombination in human embryonic stem cells using CRISPR/Cas9 nickase and a long DNA donor template. *Protein Cell* **5:** 258–260.

Shen B, Zhang W, Zhang J, Zhou J, Wang J, Chen L, Wang L, Hodgkins A, Iyer V, Huang X, et al. 2014. Efficient genome modification by CRISPR-Cas9 nickase with minimal off-target effects. *Nat Methods* **11:** 399–402.

Thomas KR, Capecchi MR. 1987. Site-directed mutagenesis by gene targeting in mouse embryo-derived stem cells. *Cell* **51:** 503–512.

Wang H, Yang H, Shivalila CS, Dawlaty MM, Cheng AW, Zhang F, Jaenisch R. 2013. One-step generation of mice carrying mutations in multiple genes by CRISPR/Cas-mediated genome engineering. *Cell* **153:** 910–918.

Yang H, Wang H, Shivalila CS, Cheng AW, Shi L, Jaenisch R. 2013. One-step generation of mice carrying reporter and conditional alleles by CRISPR/Cas-mediated genome engineering. *Cell* **154:** 1370–1379.

Yasue A, Mitsui SN, Watanabe T, Sakuma T, Oyadomari S, Yamamoto T, Noji S, Mito T, Tanaka E. 2014. Highly efficient targeted mutagenesis in one-cell mouse embryos mediated by the TALEN and CRISPR/Cas systems. *Sci Rep* **4:** 5705.

Cite this introduction as *Cold Spring Harb Protoc*; doi:10.1101/pdb.top087536

Protocol 1

Generation of Genetically Modified Mice Using the CRISPR–Cas9 Genome-Editing System

Jorge Henao-Mejia,[1,2,6,7] Adam Williams,[3,6,7] Anthony Rongvaux,[4] Judith Stein,[4,5] Cynthia Hughes,[4,5] and Richard A. Flavell[4,5,7]

[1]*Institute for Immunology, Perelman School of Medicine, University of Pennsylvania, Philadelphia, Pennsylvania 19104;* [2]*Division of Transplant Immunology, Department of Pathology and Laboratory Medicine, Children's Hospital of Philadelphia, University of Pennsylvania, Philadelphia, Pennsylvania 19104;* [3]*The Jackson Laboratory for Genomic Medicine, Department of Genetics and Genome Sciences, University of Connecticut Health Center, Farmington, Connecticut 06032;* [4]*Department of Immunobiology, Yale University School of Medicine, New Haven, Connecticut 06520;* [5]*Howard Hughes Medical Institute, Yale University School of Medicine, New Haven, Connecticut 06520*

Genetically modified mice are extremely valuable tools for studying gene function and human diseases. Although the generation of mice with specific genetic modifications through traditional methods using homologous recombination in embryonic stem cells has been invaluable in the last two decades, it is an extremely costly, time-consuming, and, in some cases, uncertain technology. The recently described CRISPR–Cas9 genome-editing technology significantly reduces the time and the cost that are required to generate genetically engineered mice, allowing scientists to test more precise and bold hypotheses in vivo. Using this revolutionary methodology we have generated more than 100 novel genetically engineered mouse strains. In the current protocol, we describe in detail the optimal conditions to generate mice carrying point mutations, chromosomal deletions, conditional alleles, fusion tags, or endogenous reporters.

MATERIALS

It is essential that you consult the appropriate Material Safety Data Sheets and your institution's Environmental Health and Safety Office for proper handling of equipment and hazardous materials used in these recipes.

RECIPES: Please see the end of this protocol for recipes indicated by <R>. Additional recipes can be found online at http://cshprotocols.cshlp.org/site/recipes.

Reagents

Agarose I (Molecular Biology Grade) (Life Technologies 17850)

Agilent RNA 6000 Pico Kit (Agilent Technologies 5067-1513)

Circular DNA donor vector plasmid (optional; see Step 2)

These vectors are used for performing large fragment insertions.

CutSmart buffer (New England Biolabs B7204S)

[6]These authors contributed equally to this work.

[7]Correspondence: jhena@mail.med.upenn.edu, adam.williams@jax.org, richard.flavell@yale.edu

Copyright © Cold Spring Harbor Laboratory Press; all rights reserved

Cite this protocol as *Cold Spring Harb Protoc*; doi:10.1101/pdb.prot090704

DNeasy Blood & Tissue Kit (QIAGEN 69581)
EDTA (0.5 M, pH 8.0) (AmericanBio AB00502-01000)
EndoFree Plasmid Maxi Kit (QIAGEN)
Gene-specific primers for genotyping (Sigma-Aldrich)
Human chorionic gonadotropin (hCG; Sigma-Aldrich C8554)
Hyaluronidase (from bovine testes) (Sigma-Aldrich H4272)
M2 medium (Sigma-Aldrich M7167)
M16 medium (Sigma-Aldrich M7292)
MEGAclear Transcription Clean-Up Kit (Life Technologies AM1908)
MEGAshortscript T7 Transcription Kit (Life Technologies AM1354)
Mice, C57BL/6, age 3–4 wk (female) and >8 wk (male) (The Jackson Laboratory)

These mice are used to produce zygotes.

Mice, ICR (Institute for Cancer Research) (CD-1) mice (males and females) (Charles Rivers Laboratories)

These are used for pseudopregnant foster mothers and vasectomized males.

Mineral oil, light (Sigma-Aldrich 330779)
mMESSAGE mMACHINE T7 Ultra Transcription Kit (Life Technologies AM1345)
Nuclease-free water (not DEPC [diethylpyrocarbonate]-treated) (Life Technologies AM9932)
Oligo DNAs used for in vitro transcription (Sigma-Aldrich)

sgRNA template: 5′-NNNNNNNNNNNNNNNNNNNNGTTTTAGAGCTAGAAATAGCAAGTTAAAATAAGGCTAGTCCGTTATCAACTTGAAAAAGTGGCACCGAGTCGGTGCTTTTTT-3′
sgRNA_R: 5′-AAAAAAGCACCGACTCGGTG-3′
T7-sgRNA_F: 5′-GAAATTAATACGACTCACTATAGGGAGANNNNNNNNNNNNNNNNNNGTTTTAGA-3′

Phenol:chloroform:isoamyl alcohol 25:24:1 (saturated with 10 mM Tris, pH 8.0, 1 mM EDTA) (Sigma-Aldrich 2069)
Platinum *Taq* DNA Polymerase, High-Fidelity (Life Technologies 11304-011)
pMJ920 plasmid (Addgene 42234)

This plasmid contains the Cas9 coding region.

Pregnant mare serum gonadotropin (PMSG; Sigma-Aldrich G4527)
QIAquick PCR Purification Kit (QIAGEN 28106)
Single-strand DNA (ssDNA) (Integrated DNA Technologies) (see Step 2)

These oligo DNAs are used to insert small DNA fragments and should be ordered as a 4-nm ultramer, desalted. Include phosphorothioate linkages to prevent degradation in the first and last three nucleotides of the oligo DNAs.

Sodium dodecyl sulfate (SDS) (20%) <R>
ssDNA/RNA Clean and Concentrator kit (Zymo Research D7010)
SURVEYOR Mutation Detection Kit - S100 (Integrated DNA Technologies 706020) (optional; see Steps 59–64)
TOPO TA Cloning Kit for subcloning (Invitrogen 450641) (optional; see Steps 59–64)
Tris (1 M, pH 7.0) (Life Technologies AM9850G)
Tris (1 M, pH 8.0) (Life Technologies AM9855G)
Tris-acetate-EDTA (TAE) buffer <R>
Tsg DNA polymerase (Lambda Biotech D101-200) (optional; see Steps 59–64)
XbaI (New England BioLabs R0145S)

Equipment

2100 Bioanalyzer (Agilent Technologies G2940CA)
C1000 Touch Thermal Cycler (Bio-Rad)

 Cite this protocol as *Cold Spring Harb Protoc*; doi:10.1101/pdb.prot090704

CO_2 incubator (Thermo Scientific BB15)
Femtotip injection capillaries, sterile (Eppendorf 930000035)
IX73 Inverted microscope, with Hoffman optics (Olympus IX73)
Microcentrifuge (Eppendorf 5424)
Microcentrifuge, refrigerated (Eppendorf 5424R)
Microinjector, CellTram vario (Eppendorf 5176000084)
Micromanipulator, three-axis hanging joystick, oil hydraulic (Narishige MMO-202ND)
Micropipettes, small holding, 20° angled (Origio MPH-SM-20)
Needles, 26 G × 3/8 in. (Becton Dickinson 309625)
Stereomicroscope (Olympus SZX7)
Tissue culture dishes, BD Falcon, polystyrene, sterile (60 × 15 mm) (Becton Dickinson 351007)

These dishes are used for embryo culture.

Tissue culture dishes, BD Falcon, polystyrene, sterile (100 × 20 mm) (Becton Dickinson 353003)

Bottoms are suitable for oocyte/embryo collection; lids are suitable for micromanipulation.

Transfer pipette capillaries, glass, thin-wall, 6 in., 1 mm o.d., 0.75 mm i.d. (WPI TW100F-6)

METHOD

Single Guide RNA (sgRNA) Design

1. For sgRNA design, follow the protocol described by Ran et al. (2013).

   The type of sgRNA required will depend on the specific type of genetic alteration desired.

   - For gene knockouts by indels, design the sgRNA for the targeted gene. For multiple-gene knockouts by indels in a single embryo, design two or more sgRNAs to target each gene.
   - For targeted chromosomal deletions, design two sgRNAs, each of which will generate a double-strand break (DSB) at the start and end point of the sequence to be deleted.
   - To introduce a small DNA sequence (e.g., point mutation, fusion tag), design a single sgRNA.
   - To generate a conditional allele by inserting two *loxP*s in *cis*, design two sgRNAs at the insertion sites of interest.
   - To introduce a large DNA sequence (e.g., a fluorescent protein) by homologous recombination (HR), design a single sgRNA at the insertion site of interest.

Donor Design

2. Design donor DNA depending on the specific type of genetic alteration desired.

   - To introduce a short DNA sequence (e.g., point mutation, fusion tag), design a single-stranded DNA (ssDNA) oligonucleotide that encodes the desired DNA insertion flanked on each side by 50–70 bases homologous to the sequence surrounding the sgRNA-mediated DSB. Insert the sequence as close as possible to the generated DSB.

     We have been able to introduce up to 100 bases with homologous regions of 50 bases on each side.

   - To generate a conditional allele by introducing two *loxP* sites, design two ssDNA oligonucleotides that encode the *loxP* sequence flanked on each side by 63 bases homologous to the sequence surrounding the sgRNA-mediated DSB. Insert the *loxP* sequence as close as possible to the generated DSB.
   - To introduce a large DNA sequence (e.g., a fluorescent protein), design and construct a plasmid that encodes the DNA sequence to be inserted flanked by >2 kb of homology on

each side using standard molecular biology methods. Insert the DNA sequence of interest as close as possible to the generated DSB.

We have been able to introduce large fragments of DNA with smaller homology regions (~500 bp).

*When introducing a small or a large DNA fragment, a potential problem should be considered. After induction of the DSB and the DNA sequence has been inserted in the genome by homology-directed repair (HDR) or HR, it is possible that the Cas9 endonuclease can recut the modified allele if the sgRNA and protospacer-adjacent motif (PAM) sequence have not been modified in the donor ssDNA or circular DNA vector. To prevent this, place the DNA sequence that encodes the desired insertion in a position in which the PAM sequence or the six nucleotides closest to the PAM sequence are disrupted (*Fig. 1C*).*

In Vitro Transcription and Purification of Cas9 mRNA

3. Linearize the pMJ920-Cas9 plasmid containing the Cas9 endonuclease coding region under the T7 promoter with XbaI by incubating the following reaction for 16 h at 37°C.

Reagent	Volume
Circular pMJ920-Cas9 plasmid	5 µL (5 µg of plasmid total)
XbaI restriction enzyme	0.5 µL
CutSmart buffer (10×)	5 µL
Nuclease-free water	39.5 µL

4. Add 1.5 µL of proteinase K (150 ng/µL) and 2.5 µL 20% SDS to the reaction mix. Incubate for 30 min at 50°C.
5. Analyze 5 µL of the digestion reaction on a 1% (w/v) agarose gel in TAE buffer to verify the successful linearization of the pMJ920-Cas9 plasmid.
6. Purify the linearized pMJ920-Cas9 plasmid with phenol:chloroform:isoamyl alcohol (25:24:1) according to the manufacturer's instructions. Resuspend the purified linearized pMJ920-Cas9 plasmid to a concentration of 1 µg/µL using nuclease-free water.
7. Use 1 µg of the purified linearized pMJ920-Cas9 plasmid as the template for in vitro transcription to generate Cas9 mRNA using the mMESSAGE mMACHINE T7 Ultra Transcription Kit according to the manufacturer's instructions.
8. Purify the Cas9 mRNA using the MEGAclear Transcription Clean-Up Kit according to the manufacturer's instructions, and elute it with 65 µL of elution buffer.

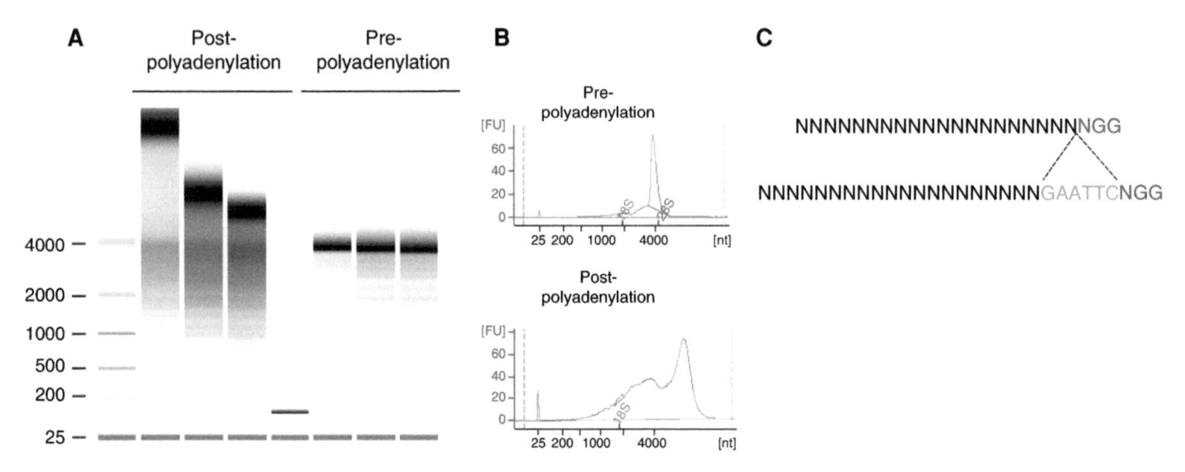

FIGURE 1. Quality control of Cas9 mRNA in a 2100 Bioanalyzer and ssDNA donor oligo design. (*A*) Electrophoresis of in vitro–transcribed Cas9 mRNA pre- and postpolyadenylation. (*B*) Electropherogram of in vitro–transcribed Cas9 mRNA pre- and postpolyadenylation. (*C*) Representation of a sequence that can be targeted with a specific sgRNA. Insertion of EcoRI (as an example; blue) adjacent to the PAM sequence (red) will preclude the Cas9 endonuclease from cutting again after the genome has been repaired by HDR.

Cite this protocol as *Cold Spring Harb Protoc*; doi:10.1101/pdb.prot090704

9. After purification, determine the quality of the Cas9 mRNA by analyzing 2.5 µL of the Cas9 mRNA pre- and postpolyadenylation in a Bioanalyzer using the Agilent RNA 6000 Pico reagents and chips according to the manufacturer's instructions.

   A discrete ~4-kb band should be observable in the sample prepolyadenylation, with an upward shift post-polyadenylation (Fig. 1A). Smeared bands indicate degradation. Discard degraded samples.

10. Calculate the concentration of the Cas9 mRNA according to the values reported in the Bioanalyzer (Fig. 1B).

11. Dilute the Cas9 mRNA to 400 ng/µL in nuclease-free microinjection buffer (5 mM Tris-HCl, pH 7.4, 0.1 mM EDTA, pH 8.0). Freeze aliquots at −80°C.

In Vitro Transcription and Purification of sgRNAs

12. For sgRNA preparation, add the T7 promoter sequence to the sgRNA template(s) (from Step 1, as appropriate) by PCR amplification using the T7-sgRNA_F and sgRNA_R primers as follows.

Reagent	Volume	Final conc.
sgRNA template (300 ng/µL)	4 µL	1200 ng
High-fidelity buffer (10×)	40 µL	1×
$MgSO_4$ (50 mM)	12 µL	1.5 mM
dNTPs (10 mM)	8 µL	0.2 mM
T7-sgRNA_F (10 µM)	8 µL	0.2 µM
sgRNA_R (10 µM)	8 µL	0.2 µM
High-fidelity polymerase	2.4 µL	
Nuclease-free water	317.6 µL	

13. Divide this PCR into eight PCR tubes (50 µL/each) and perform amplification using the following cycling conditions.

Cycle number	Denaturation	Annealing	Elongation
1	94°C, 2 min		
2–35	94°C, 20 sec	55°C, 30 sec	68°C, 40 sec
36			68°C, 7 min

14. Analyze 5 µL of the PCR on a 2.5% (w/v) agarose gel in 1× TAE buffer to verify that the product is unique and of the expected size.

   The size of the PCR product is ~120 bp for sgRNA templates.

15. Purify the T7-sgRNA PCR product using the QIAquick PCR purification kit according to the manufacturer's instructions.

16. Elute in 30 µL of nuclease-free water and resuspend the purified PCR product to a concentration of 120.5 ng/µL.

17. Use 1 µL of the purified T7-sgRNA PCR product as the template for in vitro transcription of sgRNA using the MEGAshortscript T7 Transcription Kit according to the manufacturer's instructions.

18. Purify the sgRNA using the MEGAclear Transcription Clean-Up Kit and elute with elution buffer according to the manufacturer's instructions.

19. Dilute the sgRNA to 500 ng/µL in nuclease-free microinjection buffer (5 mM Tris-HCl, pH 7.4, 0.1 mM EDTA, pH 8.0).

20. Verify the sgRNA's quality on a 2% (w/v) agarose gel in TAE buffer.

   Smeared bands indicate degradation. Discard degraded samples.

21. Freeze 30–40 µL aliquots at −80°C.

Purification of Donor DNA

For ssDNA Used to Insert Small DNA Fragments

22. Purify the donor ssDNA (from Step 2, as appropriate) using the ssDNA/RNA clean and concentrator kit according to the manufacturer's instructions and elute with 100 µL of elution buffer.
23. Dilute the donor ssDNA to 500 ng/µL in nuclease-free microinjection buffer (5 mM Tris-HCl, pH 7.4, 0.1 mM EDTA, pH 8.0).
24. Freeze 30–40 µL aliquots at −80°C.

For Circular DNA Used to Insert Large DNA Fragments

25. Prepare the circular DNA donor vector (from Step 2) using the EndoFree plasmid maxi kit according to the manufacturer's instructions.
26. Purify the circular DNA donor vector using the QIAquick PCR purification kit according to the manufacturer's instructions. Elute in 30 µL of nuclease-free microinjection buffer (5 mM Tris-HCl, pH 7.4, 0.1 mM EDTA, pH 8.0). Freeze aliquots at −80°C.

Preparation of Samples for Microinjection

27. Centrifuge thawed stock aliquots of the Cas9 mRNA (from Step 11), sgRNAs (from Step 21), and/or ssDNA donor oligos (from Step 24) at 13,200 rpm for 10 min at 4°C.
28. Depending on the aim of the experiment, prepare the appropriate injection mix by combining the components in an RNase-free microcentrifuge tube as described below.
    - For gene disruption by nonhomologous end joining or small/large chromosomal deletions add 10 µL of Cas9 mRNA stock (from Step 11), 4 µL of each sgRNA stock (from Step 21), and nuclease-free microinjection buffer up to 40 µL.

      The final concentrations are 100 ng/µL of Cas9 mRNA and 50–100 ng/µL of each sgRNA.

    - For point mutations, small tag insertions or conditional allele generation by HDR add 10 µL of Cas9 mRNA stock (from Step 11), 4 µL of each sgRNA stock (from Step 21), 8 µL of donor ssDNA stock (from Step 24), and nuclease-free microinjection buffer up to 40 µL.

      The final concentrations are 100 ng/µL of Cas9 mRNA, 50–100 ng/µL of each sgRNA, and 100 ng/µL of each donor ssDNA.

    - For large fragment insertion by HR add 10 µL of Cas9 mRNA stock (from Step 11), 4 µL of each sgRNA stock (from Step 21), the circular DNA donor vector (from Step 26), and nuclease-free microinjection buffer up to 40 µL.

      The final concentrations are 100 ng/µL of Cas9 mRNA, 50–100 ng/µL of each sgRNA, and 200 ng/µL of the circular DNA donor vector.

29. Centrifuge the microinjection mixture at 13,200 rpm for 6 min at 4°C. Remove 30 µL and transfer to a new RNase-free microcentrifuge tube.
30. Centrifuge the microinjection mixture at 13,200 rpm for 6 min at 4°C. Place the mixture on ice and begin microinjection.

Zygote Preparation

31. Inject 15 female C57BL/6-(3- to 4-wk-old) mice with PMSG (5 IU) on day 1.
32. After 48 h (i.e., on day 3), inject female mice with hCG (5 IU). After the hCG injection, house female mice with C57BL/6 male mice of proven fertility overnight.
33. On the morning of day 4, collect female mice with cervical plugs for zygote preparation. Euthanize the female mice without plugs.

 Cite this protocol as *Cold Spring Harb Protoc*; doi:10.1101/pdb.prot090704

34. Prepare the medium for embryo culture.
    i. Place several drops (30–50 µL per drop) of M16 medium on a 60-mm dish.
    ii. Cover the drops with light mineral oil.
    iii. Place the dish into a 37°C incubator for at least 30 min before use.
35. At 20–21 h after hCG injection, kill the mice and collect zygote-cumulus complexes from the oviduct.
36. Prepare the medium for embryo collection by adding hyaluronidase to M2 media to obtain a working concentration of 300 µg/mL (M2 + Hy medium).
37. Move the zygote-cumulus complexes into M2 + Hy medium in a 100-mm dish. Pipette up and down several times. Aspirate the embryos with a transfer pipette.
38. Wash several times in M2 medium.
39. Place the embryos into M16 medium at 37°C in a 5% CO_2 incubator until ready for microinjection.

Microinjection of Zygotes

40. Use a transfer pipette to transfer a group of fertilized oocytes into the injection chamber containing M16 medium. Wait for at least 5 min before starting the injection.

    The number of zygotes to be moved into the microinjection drop should be determined by the skills of the injector and quality of the setup.
41. Examine the zygotes under high power, making sure that two pronuclei are visible and that the morphology is good. Discard all zygotes that appear abnormal.
42. Aspirate the injection mix into microinjection capillary.
43. Determine whether the microinjection capillary is open and unclogged by placing the tip of the microinjection capillary close to a zygote in the same horizontal plane under a continuous flow stream.

    If the microinjection capillary is open, a stream of DNA will move the zygote away from the tip of the microinjection capillary. See Troubleshooting.
44. To prepare a zygote for injection, place the tip of the holding pipette next to the zygote, and apply a negative pressure to the pressure control unit.
45. Focus the microscope to locate the pronuclei.

    The pronucleus should also be as close as possible to the central axis of the holding pipette.
46. Refocus on the pronucleus to be injected. Bring the tip of the microinjection capillary into the same focal plane as the mid-plane of the pronucleus.
47. Move the injection pipette to the same *y*-axis position as the targeted pronucleus.
48. Adjust the height of the pipette so that the tip of the microinjection capillary appears completely sharp.
49. Move the injection pipette to a 3 o'clock position without changing its height.
50. Push the microinjection pipette through the zona pellucida into the cytoplasm and toward the pronucleus.
51. When the tip of the microinjection capillary appears to be inside the cytoplasm apply injection pressure through the injector.

    If the cytoplasm swells visibly, it has been successfully injected. If the cytoplasm does not swell, the pipette has become clogged or has not punctured the oocyte plasma membrane. If a small round "bubble" forms around the tip of the pipette, then the pipette has not punctured the plasma membrane. Cytoplasmic granules flowing out of the oocyte after removal of the injection pipette are a clear sign that the zygote will soon lyse. In this case, the oocyte should be discarded. If the zygote appears to be intact and successfully injected, it should be sorted and another zygote should be picked up for injection.

52. Quickly pull the pipette out of the zygote.
53. Repeat Steps 44–52 with the remaining zygotes.

    Some injected zygotes will inevitably lyse because of the mechanical damage caused by the injection procedure. The lysed zygotes can be distinguished from healthy ones by the appearance of the zona pellucida as it is more translucid. Typically, ~75% of the zygotes survive the microinjection.

54. After all the zygotes in the chamber have been injected immediately transfer them back into M16 medium and incubate at 37°C in a 5% CO_2 incubator.

Embryo Transfer and Production of Mice

55. Prepare pseudopregnant foster mothers by mating estrous CD1 female mice with vasectomized male mice on the same day as injection.
56. Transfer approximately 25 microinjected zygotes into oviducts of 0.5 d postcoitum (dpc) recipients.

    Alternatively, injected zygotes can be cultured in M16 medium at 37°C in a 5% CO_2 incubator until they reach the two-cell stage 24 h later and then be transferred into oviducts of 0.5 d postcoitum (dpc) recipients.

57. Deliver pups from recipient mothers at 19.5 dpc.
58. Separate male and female offspring into individual cages at 3 wk after birth.

Genotyping

59. Extract genomic DNA from tail biopsies of 10-d-old mice using a DNeasy blood and tissue kit according to the manufacturer's protocol.

For Small Genomic Modifications

60. To genotype small genomic modifications such as point mutations, gene disruptions, small sequence insertions, or conditional alleles, perform PCR amplification using gene-specific primers under the following conditions.

Reagent	Volume	Final concentration
Genomic DNA	2.5 µL	
Tsg buffer (10×)	2.5 µL	1×
$MgCl_2$	1.5 µL	1.5 mM
dNTPs (10 mM)	0.5 µL	0.2 mM
Primer_F (10 µM)	0.5 µL	0.2 µM
Primer_R (10 µM)	0.5 µL	0.2 µM
Tsg polymerase	0.3 µL	
Nuclease-free water	16.7 µL	

Cycle number	Denaturation	Annealing	Elongation
1	95°C, 2 min		
2–35	94°C, 30 sec	72°C, 30 sec, −0.5°C/cycle	72°C, 45 sec
36			72°C, 7 min

61. If desired, purify the PCR products by using the QIAquick PCR Purification Kit according to the manufacturer's instructions. Then analyze the PCR products by one of the following methods.

    - To check for indels or small mutations perform a SURVEYOR assay according to the manufacturer's instructions.

      This can be performed directly on the (unpurified) PCR product.

Cite this protocol as *Cold Spring Harb Protoc*; doi:10.1101/pdb.prot090704

- If a restriction enzyme recognition site has been inserted or removed, perform restriction enzyme digestion of the PCR product.
- To verify mutations, clone the PCR products using the TOPO TA Cloning kit and sequence according to the manufacturer's instructions.

For Insertions of Tags, loxPs*, or Deletions*

62. To genotype for insertions of tags, *loxPs* or deletions, perform PCR amplification using gene-specific primers under the following conditions.

Reagent	Volume	Final concentration
Genomic DNA	2.5 µL	
Tsg buffer (10×)	2.5 µL	1×
$MgCl_2$	1.5 µL	1.5 mM
dNTPs (10 mM)	0.5 µL	0.2 mM
Primer_F (10 µM)	0.5 µL	0.2 µM
Primer_R (10 µM)	0.5 µL	0.2 µM
Tsg polymerase	0.3 µL	
Nuclease-free water	16.7 µL	

Cycle number	Denaturation	Annealing	Elongation
1	95°C, 2 min		
2–35	94°C, 30 sec	72°C, 30 sec, −0.5°C/cycle	72°C, 45 sec
36			72°C, 7 min

63. Analyze the PCR product on a 2% (w/v) agarose gel in 1X TAE buffer to verify that the product is unique and of the expected size.

For Large Insertions

64. To genotype for large insertions perform PCR amplification for large fragments using gene-specific primers that allows detecting the insertion of the fragment at the specific genomic location.

    PCR conditions should be optimized in each case.

TROUBLESHOOTING

Problem (Step 43): The microinjection pipette is closed or clogged.
Solution: Flush DNA with high power through the microinjection capillary by using the "Clear" function on the microinjector. Repeat the test. If the tip is still not open, tip it carefully on the holding pipette to break the tip to a larger diameter. If the diameter becomes too large, or the tip is still not open, discard the pipette and use a new one.

DISCUSSION

Based on the targeting experiments we have performed (more than 400 different microinjections producing more than 100 new mouse strains), the great majority of microinjections with only Cas9 mRNA and sgRNAs resulted in mutant alleles containing indels with high efficiency (>80% of pups were targeted for one allele). The efficiency of targeted chromosomal deletions greatly depends on the size of the deletion; we have successfully deleted chromosomal regions up to ~200 kb and have observed efficiencies for targeted deletions that range from 10% to 40% of born pups being targeted. With co-injection of donor ssDNA, the efficiency of HDR varies from 10% to 60% depending on the size of the fragment that is being inserted/modified. Using double-stranded plasmid donor DNA, the efficiency of HR varies greatly and requires significant optimization.

RELATED INFORMATION

A more detailed discussion on the background and uses of these techniques is available in Introduction: Editing the Mouse Genome Using the CRISPR–Cas9 System (Williams et al. 2016).

RECIPES

SDS

Also called sodium dodecyl sulfate or sodium lauryl sulfate. To prepare a 20% (w/v) solution, dissolve 200 g of electrophoresis-grade SDS in 900 mL of H_2O. Heat to 68°C and stir with a magnetic stirrer to assist dissolution. If necessary, adjust the pH to 7.2 by adding a few drops of concentrated HCl. Adjust the volume to 1 L with H_2O. Store at room temperature. Sterilization is not necessary. Do not autoclave.

TAE

Prepare a 50× stock solution in 1 L of H_2O:
242 g of Tris base
57.1 mL of acetic acid (glacial)
100 mL of 0.5 M EDTA (pH 8.0)

The 1× working solution is 40 mM Tris-acetate/1 mM EDTA.

REFERENCES

Ran FA, Hsu PD, Wright J, Agarwala V, Scott DA, Zhang F. 2013. Genome engineering using the CRISPR–Cas9 system. *Nat Protoc* **8:** 2281–2308.

Williams A, Henao-Mejia J, Flavell R. 2016. Editing the mouse genome using the CRISPR–Cas9 system. *Cold Spring Harb Protoc* doi: 10.1101/pdb.top087536.

Cite this protocol as *Cold Spring Harb Protoc*; doi:10.1101/pdb.prot090704

CHAPTER 11

Genome Editing in Human Pluripotent Stem Cells

Cory Smith,[1,2,3] Zhaohui Ye,[1,2] and Linzhao Cheng[1,2,3,4]

[1]*Division of Hematology, Department of Medicine, Johns Hopkins University School of Medicine, Baltimore, Maryland 21205;* [2]*Stem Cell Program, Institute for Cell Engineering, Johns Hopkins University School of Medicine, Baltimore, Maryland 21205;* [3]*Predoctoral Training Program in Human Genetics, Johns Hopkins University School of Medicine, Baltimore, Maryland 21205*

Pluripotent stem cells (PSCs), defined by their capacity for self-renewal and differentiation into all cell types, are an integral tool for basic biological research and disease modeling. However, full use of PSCs for research and regenerative medicine requires the ability to precisely edit their DNA to correct disease-causing mutations and for functional analysis of genetic variations. Recent advances in DNA editing of human stem cells (including PSCs) have benefited from the use of designer nucleases capable of making double-strand breaks (DSBs) at specific sequences that stimulate endogenous DNA repair. The clustered, regularly interspaced short palindromic repeats (CRISPR)–Cas9 system has become the preferred designer nuclease for genome editing in human PSCs and other cell types. Here we describe the principles for designing a single guide RNA to uniquely target a gene of interest and describe strategies for disrupting, inserting, or replacing a specific DNA sequence in human PSCs. The improvements in efficiency and ease provided by these techniques allow individuals to precisely engineer PSCs in a way previously limited to large institutes and core facilities.

EMBRYONIC AND PLURIPOTENT STEM CELLS

Embryonic stem cells (ESCs) are derived from the inner cell mass of a developing blastocyst and can be expanded in vitro for decades without losing their full potential for differentiation and self-renewal. ESCs hold great potential for basic biological research, disease modeling, and as a cell source for regenerative medicine. However, because of the destructive nature of their derivation, an autologous source of ESCs is not available without therapeutic cloning. Also, the high monetary and ethical costs associated with obtaining oocytes limit the practical applications and scalability of their use. The discovery that four simple transcription factors could reprogram somatic fibroblasts to an embryonic-like state (i.e., induced PSCs) provides an autologous cell source for individualized regenerative medicine and basic research, including disease modeling, high-throughput drug screening, and access to human cell types previously unavailable for live study, including neurons and astrocytes. The full realization of these PSCs will rely on the ability to precisely manipulate their DNA to investigate the effects of these variants in isolation or to cure a disease-causing variant ex vivo before cell-based regenerative therapy.

Gene targeting using a double-stranded DNA donor with long homology arms (1–4 kb) is a well-established technique across many animal models and cell lines. Although this approach has been used in mouse PSCs, the many technical limitations using human PSCs prevented successful modification of this technique: Single cells rarely survive without cell-to-cell contacts and signals that aid in the formation of the early colony after the stress of electroporation, and the rates of homologous recom-

[4]Correspondence: lcheng2@jhmi.edu

Copyright © Cold Spring Harbor Laboratory Press; all rights reserved

Cite this introduction as *Cold Spring Harb Protoc;* doi:10.1101/pdb.top086819

bination are significantly lower in human PSCs than in mice. The first limitation was addressed by the development of Rho-associated kinase inhibitor that dramatically increases survival and proper colony formation after the single-cell digestion of human PSCs required for electroporation; the second issue was addressed by the development of sequence-specific nucleases that generate a DSB in DNA at a targeted location, increasing the efficiency of recombination at that site by several orders of magnitude (Hockemeyer et al. 2009; Zou et al. 2009). DSBs are detected by the cell's endogenous DNA repair machinery, which proceeds through either the error-prone nonhomologous end joining (NHEJ) pathway that often results in small indel mutations around the break site or by homology-dependent repair (HDR) using a DNA donor template. These tools allow researchers to both disrupt genes through NHEJ and to create single-nucleotide variants with an HDR donor to determine its effects in isolation when compared with an otherwise isogenic cell line.

CRISPR–Cas EDITING OF PSCs

The *Streptococcus pyogenes*–derived adaptive bacterial immune system CRISPR–Cas degrades foreign DNA in a sequence-specific manner guided by short guide RNAs (gRNAs) to induce a DSB. In particular, the type II CRISPR system only requires a single protein component, Cas9, and a single gRNA to determine its target and cleave DNA. This system has been successfully adapted for mammalian expression (Cong et al. 2013; Mali et al. 2013). A major advantage of CRISPR–Cas9 is the ability to rapidly design, synthesize, assemble, and test gRNAs targeting new genes or noncoding regions of interest, as well as the ability to multiplex by transfecting additional gRNAs. Several studies—largely conducted using cancer cell lines (e.g., HEK 293T or U20S)—revealed higher than expected off-target mutagenesis, including some sites that were altered more frequently than the intended target and other sites that included up to five mismatches compared with the gRNA (Fu et al. 2013; Hsu et al. 2013; Mali et al. 2013); further screening in animal embryos and human stem cells reported minimal off-target effects (Wang et al. 2013; Niu et al. 2014; Wu et al. 2014). Several complementary approaches to screen human PSCs for potential off-target mutagenesis such as targeted sequencing of top sites predicted most likely to cleave based on sequence similarity revealed high specificity of these gRNAs tested in human PSCs relative to HEK 293Ts (Mali et al. 2013; Smith et al. 2014; Li et al. 2015). Additionally, whole-genome sequencing conducted on CRISPR–Cas9-targeted PSCs revealed new mutations in each clone, but none were similar to the gRNA target site or recurrent between the clones, indicating that genome editing of PSCs can be performed with minimal off-target mutagenesis genome-wide (Smith et al. 2014; Suzuki et al. 2014; Veres et al. 2014; Yang et al. 2014).

There are many approaches for genome editing to either knock out a gene, induce a specific single-nucleotide change, or insert a larger fragment for transgene expression or to add a fluorescent tag (Fig. 1). The most basic strategy involves a single guide RNA targeting the protein coding sequence of a gene to disrupt expression by a frameshift mutation, often leading to degradation through nonsense-mediated decay. If two gRNAs are used in close proximity (~20 bp–10 kb) the intervening sequences is deleted, often with precise breakpoints 3 nucleotides upstream of the NGG, although small indels are observed around the junction at a low frequency. This technique can be used to delete exons that could not otherwise be targeted because of highly conserved domains and also provides a tool to excise defined regions, making it a promising approach to investigate noncoding variants such as regulatory elements over a broad range of deletion sizes. To induce specific mutations (e.g., single-nucleotide changes), the classic approach uses a double-stranded DNA donor plasmid with homology arms of ~0.5–2 kb. Using CRISPR–Cas9, this approach can produce dramatically higher efficiencies (~10%) without using drug selection (Byrne et al. 2015) and can also be used to insert transgenes into a predetermined safe harbor locus (e.g., the adeno-associated virus integration site 1) or to tag a lineage-specific gene to monitor expression during differentiation. Another targeting approach uses a single-stranded DNA oligonucleotide (ssODN) of ~70–110 bases, although 90 bp was optimal in

 Cite this introduction as *Cold Spring Harb Protoc*; doi:10.1101/pdb.top086819

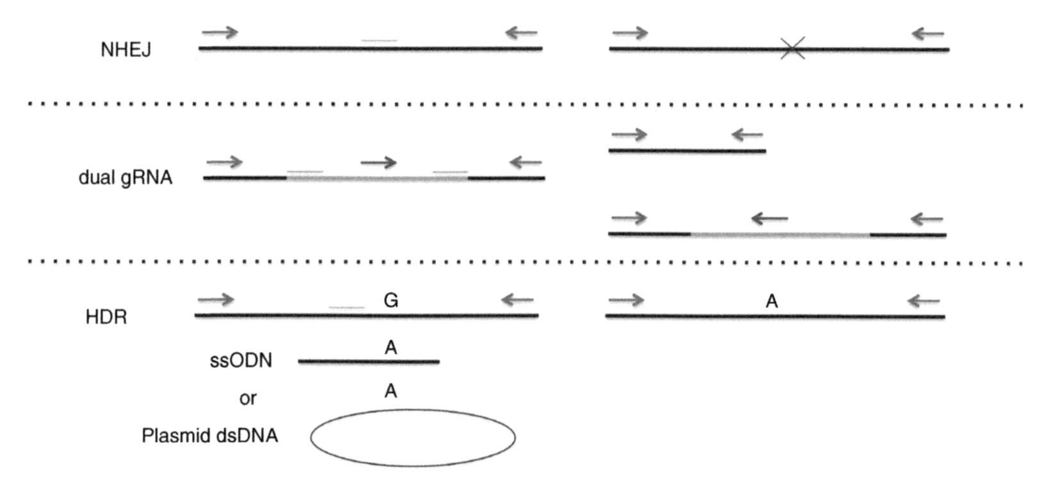

FIGURE 1. Targeting approaches for genome editing. NHEJ uses a single guide RNA (green) targeting the protein coding sequence of a gene to disrupt expression by a frameshift mutation (×). Dual guide RNA uses two gRNAs in close proximity to delete the intervening sequences (blue arrow). HDR can be used with either a single-stranded DNA oligonucleotide (ssODN) or a double-stranded DNA donor (plasmid dsDNA) to induce specific mutations.

human PSCs (Yang et al. 2013). The same study also reported that it was ideal to use the strand complementary to the gRNA for the ssODN as opposed to the same strand. It is best to design the gRNA as close to the mutation as possible (ideally, within 10 nucleotides for ssODNs or 200 nucleotides for plasmid donors) (Xie et al. 2014).

In Protocol 1: A Method for Genome Editing in Human Pluripotent Stem Cells (Smith et al. 2016), we describe a method for genome editing in PSCs with CRISPR–Cas9 to disrupt or delete a predetermined DNA sequence or to alter a nucleotide of interest for further study. Targeting considerations are discussed to design gRNAs for both knockout and knock-in strategies that have the least chance of off-target mutagenesis predicted bioinformatically and tested experimentally in the PSC line of interest. We also provide a simplified procedure for gRNA synthesis and assembly and discuss considerations for transfection-quality DNA needed to modify the traditionally difficult-to-transfect PSCs. Briefly, transfection is achieved through electroporation using the Nucleofector 4D (Lonza), although similar efficiencies in PSCs can be obtained using other technologies (e.g., Life Technologies' Neon Transfection System). With initially ~1% NHEJ disruption efficiencies for Cas9 in human PSCs (Mali et al. 2013) clonal isolation and screening for targeted clones remains the most time-consuming and costly part of the procedure as hundreds to thousands of clones need to be screened to find several with the intended mutation. Although feasible, several improvements have been made to increase the efficiency of editing by using either fluorescence-assisted cell sorting for Cas9_GFP$^+$ cells (Ding et al. 2013) or by integrating a dox-inducible Cas9 system (González et al. 2014). The future application of these genome-editing techniques to PSCs will allow cell-based functional investigations of any genetic variant of interest such as the incoming torrent of genome-wide association data or potential causative variants in rare case anomalies at the extremes of phenotypes.

REFERENCES

Byrne SM, Ortiz L, Mali P, Aach J, Church GM. 2015. Multi-kilobase homozygous targeted gene replacement in human induced pluripotent stem cells. *Nucleic Acids Res* **43:** e21.

Cong L, Ran FA, Cox D, Lin S, Barretto R, Habib N, Hsu PD, Wu X, Jiang W, Marraffini LA, et al. 2013. Multiplex genome engineering using CRISPR/Cas systems. *Science* **339:** 819–823.

Ding Q, Regan SN, Xia Y, Oostrom LA, Cowan CA, Musunuru K. 2013. Enhanced efficiency of human pluripotent stem cell genome editing through replacing TALENs with CRISPRs. *Cell Stem Cell* **12:** 393–394.

Fu Y, Foden JA, Khayter C, Maeder ML, Reyon D, Joung JK, Sander JD. 2013. High-frequency off-target mutagenesis induced by CRISPR–Cas nucleases in human cells. *Nat Biotechnol* **31:** 822–826.

González F, Zhu Z, Shi ZD, Lelli K, Verma N, Li QV, Huangfu D. 2014. An iCRISPR platform for rapid, multiplexable, and inducible genome editing in human pluripotent stem cells. *Cell Stem Cell* **15:** 215–226.

Hockemeyer D, Soldner F, Beard C, Gao Q, Mitalipova M, DeKelver RC, Katibah GE, Amora R, Boydston EA, Zeitler B, et al. 2009. Efficient targeting of expressed and silent genes in human ESCs and iPSCs using zinc-finger nucleases. *Nat Biotechnol* **27:** 851–857.

Hsu PD, Scott DA, Weinstein JA, Ran FA, Konermann S, Agarwala V, Li Y, Fine EJ, Wu X, Shalem O, et al. 2013. DNA targeting specificity of RNA-guided Cas9 nucleases. *Nat Biotechnol* **31:** 827–832.

Li HL, Fujimoto N, Sasakawa N, Shirai S, Ohkame T, Sakuma T, Tanaka M, Amano N, Watanabe A, Sakurai H, et al. 2015. Precise correction of the dystrophin gene in Duchenne muscular dystrophy patient induced pluripotent stem cells by TALEN and CRISPR–Cas9. *Stem Cell Rep* **4:** 143–154.

Mali P, Yang L, Esvelt KM, Aach J, Guell M, DiCarlo JE, Norville JE, Church GM. 2013. RNA-guided human genome engineering via Cas9. *Science* **339:** 823–826.

Niu Y, Shen B, Cui Y, Chen Y, Wang J, Wang L, Kang Y, Zhao X, Si W, Li W, et al. 2014. Generation of gene-modified cynomolgus monkey via Cas9/RNA-mediated gene targeting in one-cell embryos. *Cell* **156:** 836–843.

Smith C, Gore A, Yan W, Abalde-Atristain L, Li Z, He C, Wang Y, Brodsky RA, Zhang K, Cheng L, et al. 2014. Whole-genome sequencing analysis reveals high specificity of CRISPR/Cas9 and TALEN-based genome editing in human iPSCs. *Cell Stem Cell* **15:** 12–13.

Smith C, Ye Z, Cheng L. 2016. A method for genome editing in human pluripotent stem cells. *Cold Spring Harb Protoc* doi: 10.1101/pdb.prot090217.

Suzuki K, Yu C, Qu J, Li M, Yao X, Yuan T, Goebl A, Tang S, Ren R, Aizawa E, et al. 2014. Targeted gene correction minimally impacts whole-genome mutational load in human-disease-specific induced pluripotent stem cell clones. *Cell Stem Cell* **15:** 31–36.

Veres A, Gosis BS, Ding Q, Collins R, Ragavendran A, Brand H, Erdin S, Cowan CA, Talkowski ME, Musunuru K. 2014. Low incidence of off-target mutations in individual CRISPR–Cas9 and TALEN targeted human stem cell clones detected by whole-genome sequencing. *Cell Stem Cell* **15:** 27–30.

Wang H, Yang H, Shivalila CS, Dawlaty MM, Cheng AW, Zhang F, Jaenisch R. 2013. One-step generation of mice carrying mutations in multiple genes by CRISPR/Cas-mediated genome engineering. *Cell* **153:** 910–918.

Wu X, Scott DA, Kriz AJ, Chiu AC, Hsu PD, Dadon DB, Cheng AW, Trevino AE, Konermann S, Chen S, et al. 2014. Genome-wide binding of the CRISPR endonuclease Cas9 in mammalian cells. *Nat Biotechnol* **32:** 670–676.

Xie F, Ye L, Chang JC, Beyer AI, Wang J, Muench MO, Kan YW. 2014. Seamless gene correction of β-thalassemia mutations in patient-specific iPSCs using CRISPR/Cas9 and piggyBac. *Genome Res* **24:** 1526–1533.

Yang L, Guell M, Byrne S, Yang JL, De Los Angeles A, Mali P, Aach J, Kim-Kiselak C, Briggs AW, Rios X, et al. 2013. Optimization of scarless human stem cell genome editing. *Nucleic Acids Res* **41:** 9049–9061.

Yang L, Grishin D, Wang G, Aach J, Zhang CZ, Chari R, Homsy J, Cai X, Zhao Y, Fan JB, et al. 2014. Targeted and genome-wide sequencing reveal single nucleotide variations impacting specificity of Cas9 in human stem cells. *Nat Commun* **5:** 5507.

Zou J, Maeder ML, Mali P, Pruett-Miller SM, Thibodeau-Beganny S, Chou BK, Chen G, Ye Z, Park IH, Daley GQ, et al. 2009. Gene targeting of a disease-related gene in human induced pluripotent stem and embryonic stem cells. *Cell Stem Cell* **5:** 97–110.

Cite this introduction as *Cold Spring Harb Protoc*; doi:10.1101/pdb.top086819

Protocol 1

A Method for Genome Editing in Human Pluripotent Stem Cells

Cory Smith,[1,2,3] Zhaohui Ye,[1,2] and Linzhao Cheng[1,2,3,4]

[1]*Division of Hematology, Department of Medicine, Johns Hopkins University School of Medicine, Baltimore, Maryland 21205;* [2]*Stem Cell Program, Institute for Cell Engineering, Johns Hopkins University School of Medicine, Baltimore, Maryland 21205;* [3]*Predoctoral Training Program in Human Genetics, Johns Hopkins University School of Medicine, Baltimore, Maryland 21205*

Human pluripotent stem cells (PSCs) hold great potential for regenerative medicine and currently are being used as a research tool for basic discovery and disease modeling. To evaluate the role of a single genetic variant, a system of genome editing is needed to precisely mutate any desired DNA sequence in isolation and measure its effect on phenotype when compared to the isogenic parental PSC from which it was derived. This protocol describes the general targeting schemes used by researchers to edit PSCs to knock out, knock-in, or precisely alter a single nucleotide, covering conditions for electroporation, clonal isolation, and screening of edited PSCs for the targeted mutation. These recent advances simplify the procedure for genome editing, allowing individual researchers to induce nearly any desired mutation to further study its function or to reverse a disease-causing variant for future applications in regenerative medicine.

MATERIALS

It is essential that you consult the appropriate Material Safety Data Sheets and your institution's Environmental Health and Safety Office for proper handling of equipment and hazardous materials used in these recipes.

RECIPE: Please see the end of this protocol for recipes indicated by <R>. Additional recipes can be found online at http://cshprotocols.cshlp.org/site/recipes.

Reagents

Accutase (Sigma-Aldrich)
Cas9 expression vector (e.g., Cas9_GFP, Addgene 44719)
Circular DNA donor plasmid (optional; see Step 7)
Essential 8 medium (Life Technologies)
gRNA_Cloning vector (Addgene 41824)
HEK293T cells (optional; ATCC CRL-3216; see Step 3)
High-fidelity polymerase
Induced pluripotent stem cells, human
P3 Primary Cell 4D-Nucleofector X kit (Lonza)
PCR purification kit
Phosphate-buffered saline (PBS) <R>
Plasmid Maxi kit (e.g., QIAGEN)

[4]Correspondence: lcheng2@jhmi.edu

Copyright © Cold Spring Harbor Laboratory Press; all rights reserved
Cite this protocol as *Cold Spring Harb Protoc*; doi:10.1101/pdb.prot090217

Puromycin (optional; see Step 19.ii)
Single-stranded oligo deoxy-nucleotide (ssODN) (optional; see Step 7)
TOPO Cloning kit (Thermo Fischer Scientific)
Vitronectin, human, recombinant (rVTN) (Life Technologies)
Y-27632 Rho kinase inhibitor

Equipment

Benchtop centrifuge
Bioinformatics tools for evaluating gRNAs (see Step 2)
Cell strainer (40 µm)
DNA electrophoresis system
Fluorescence-activated cell sorting system (optional; see Steps 20–31)
Fluorescence microscope (optional; see Step 19)
Hemocytometer
Incubator
Lonza 4D-Nucleofector system
Thermocycler
Tissue culture plates (6-, 12-, and 24-well)
Tubes (conical, 15- and 50-mL)
Tubes (microcentrifuge, 1.5-mL)

METHOD

gRNA Design, Synthesis, and Assembly

1. (Optional) Polymerase chain reaction (PCR)-amplify ~200–1000 bp surrounding the target in the PSC of interest. Sequence the region to identify any polymorphisms that should be included in design of gRNAs or any homology donors to be used.
2. Identify and evaluate potential gRNAs using available bioinformatics tools.

   Currently available online tools include http://crispr.mit.edu/, http://www.e-crisp.org/E-CRISP/, and http://arep.med.harvard.edu/CasFinder/.

3. Select two or more gRNAs to synthesize. Evaluate experimentally using previously described protocols (Mali et al. 2013; Yang et al. 2014).

   Optionally, test the candidate construct(s) in HEK293T to identify the most efficient and specific gRNAs as well as to optimize the screening process before proceeding with human PSCs.

Transfection of Human PSCs

This protocol uses feeder-free conditions with Essential 8 medium for culturing human PSCs, although other cell culture systems that can effectively maintain pluripotency and facilitate clonal expansion can also be used. Likewise, although this protocol uses the 4D-Nucleofector X (Lonza) for transfection of Cas9 and gRNA constructs, others have reported equivalent transfection efficiencies using the Neon Transfection system (Life Technologies). Procedures are presented for both fluorescent-tag-based (Fig. 1A) and antibiotic (Fig. 1B) selection of transfected cells.

4. Three to five days before nucleofection, seed $0.5–2 \times 10^5$ PSCs into rVTN-coated plates with Essential 8 medium.
5. Change the media daily until nucleofection.
6. On the day of transfection, equilibrate a separate set of rVTN-coated six-well tissue culture plates with PBS for 1 h at 37°C.
7. Prepare the nucleofection solution in a 1.5-mL tube depending on the specific type of genetic alteration desired. Place on ice until needed.

Cite this protocol as *Cold Spring Harb Protoc*; doi:10.1101/pdb.prot090217

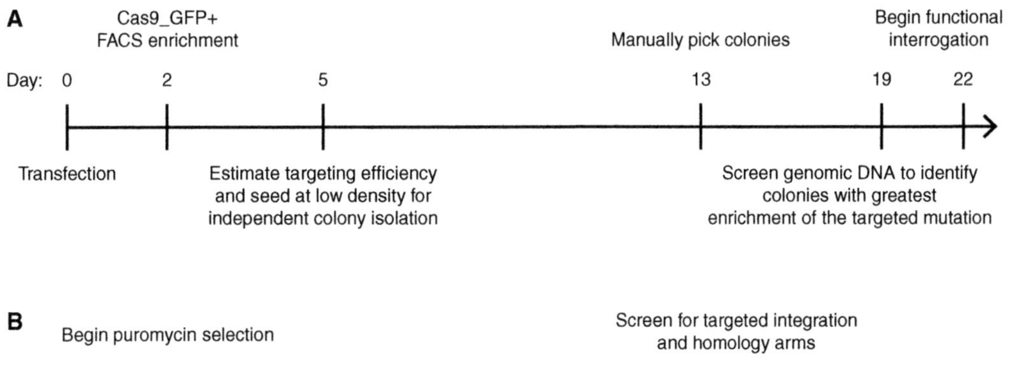

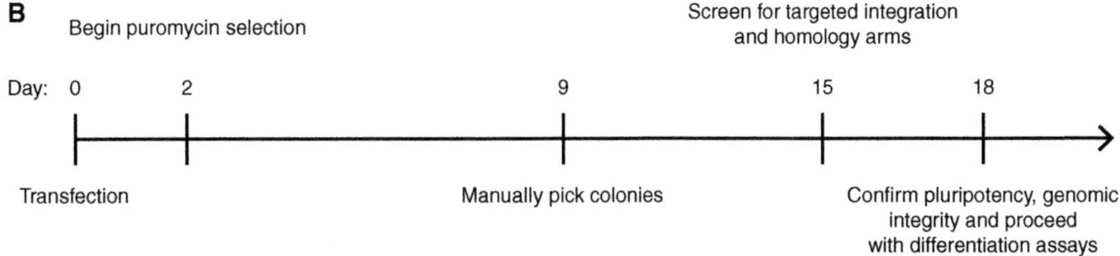

FIGURE 1. Timelines of genome editing in human PSCs. (*A*) Protocol for fluorescence-activated cell sorting selection of cells transfected with a GFP tag. (*B*) Protocol for drug selection of transfected cells.

- For nonhomologous end joining (NHEJ), mix 6 µg Cas9_GFP, 6 µg gRNA expression vector, and 100 µL Nucleofector solution.

  Transfect a control population without Cas9_GFP for FACS selection (see Step 23).

- For large deletions or multiplexing, add equal amounts of each gRNA (to a total of 6 µg) with 6 µg Cas and 100 µL Nucleofector solution.
- For homology-directed repair (HDR), include 5 µg of plasmid homologous donor or 1 nmol of ssODN.

8. Aspirate the PBS from the rVTN-coated plates (from Step 6). Replace with 2.5 mL of E8 + ROCK (to a final concentration of 10 µM). Place in the incubator at 37°C until needed.
9. Digest the cultured cells (from Step 4) to single cells using Accutase for 3 min.

   It is important to not overdigest the cells as this will result in poor survival and transfection efficiency.

10. Rinse the cells with PBS:Accutase (2:1) and pipette up and down to dissociate the cells. Transfer the cells to a 15-mL conical tube.
11. Centrifuge at 300*g* for 5 min at 4°C. Resuspend the cells in PBS.
12. Determine cell concentration using a hemocytometer.
13. Add 2×10^6 cells to 3 mL of PBS in a 15-mL tube. Centrifuge at 300*g* for 5 min at 4°C.
14. Resuspend the cells in the chilled DNA-nucleofection solution (from Step 7). Incubate on ice for 10 min.
15. Transfer the resuspended cells and DNA-nucleofection solution into an electroporation cuvette. Nucleofect using the 4D-Nucleofector using the human ES H9 program according to the manufacturer's instructions.

    See Troubleshooting.

16. Immediately add 500 µL of E8 medium to each sample in a slow dropwise manner to prevent shocking the cells.
17. Seed the transfected cells into rVTN-coated culture plates (from Step 8) as appropriate for the experimental conditions:

- For fluorescence selection, seed the transfected cells evenly between two rVTN-coated wells with the manufacturer-provided transfer pipette to a final volume of 3.2 mL in each well.
- For drug selection knock-in strategies, seed evenly into two six-well plates (12 wells total).

18. Change the media with fresh Essential 8 medium the following morning.
19. Confirm transfection efficiency:

    - For fluorescence selection, examine cells under a fluorescence microscope to determine the emergence of GFP^+ cells.

      Proceed to Step 20 to enrich cells by fluorescence-activated cell sorting (FACS).

    - For knock-in strategies, begin drug selection (e.g., with 0.5 µg/mL puromycin) at day 2 to kill cells without the marker.

      Colonies will begin to emerge after ~3 d of selection and can be picked for clonal isolation after 7 d of drug selection. Continue protocol at Step 32 for picking and expanding colonies.

FACS Enrichment of Cas9_GFP^+ PSCs

20. Equilibrate one rVTN-coated six-well tissue culture plate with E8 medium at 37°C per sample to be sorted.
21. At day 2 posttransfection, harvest the transfected PSCs with Accutase. Filter to remove anything >40 µm.
22. Resuspend in PBS and place cells on ice.
23. Use the control population transfected without Cas9_GFP to set the gate of the FACS system.
24. Collect GFP^+ cells in E8 media.
25. Wash the GFP^+-enriched cell population with PBS. Centrifuge at 300*g* for 5 min. Resuspend in E8 medium.
26. Use the live cell number determined by FACS sorting to calculate the cell concentration and determine the volume required for each sample.
27. Plate six wells of a six-well plate (from Step 20) at a range of concentrations to minimize the number of input cells contributing to each colony (50–400 cells/well).
28. Expand the low-density seeded cells by changing the medium every day.

    Early colonies can be observed at 1 d postseeding and form a clear colony morphology at days 3–5.

29. Plate 2×10^5 cells (or all remaining cells) of the GFP^+-enriched population into a single well of an rVTN-coated six-well plate as bulk culture for future screening.
30. Pellet the remainder of the GFP^+-enriched cells for genomic DNA isolation and screening for the targeted mutation.
31. Based on the overall efficiency in the bulk GFP^+ population determine the number of colonies to be picked to likely isolate several independent PSCs with the desired mutation.

Colony Expansion

32. Isolate and pick colonies manually, detaching them from the surface with a glass Pasteur pipette. Collect each colony with a 200-µL pipette and transfer to a single well of a 24-well rVTN-coated plate.

    - For fluorescently labeled cells, pick colonies at day 13 posttransfection (i.e., day 8 postplating).
    - For drug-selected samples, pick colonies at day 9 posttransfection (i.e., day 7 post–drug selection).

      See Troubleshooting.

Cite this protocol as *Cold Spring Harb Protoc*; doi:10.1101/pdb.prot090217

33. After 3–4 d of growth passage the colonies using Accutase. Transfer all the cells into a single well of a 12-well plate.
34. Once the 12-well plates are confluent (~$0.8–1.5 \times 10^6$ cells) passage using Accutase 1:10 into a single well of a six-well plate. Pellet the remaining 90% of each sample for genomic DNA extraction and screening for the targeted mutation.

Screening for Targeted Mutations

35. Extract genomic DNA (from Step 34). Use 100 ng as a template to PCR-amplify the target region to genotype for the intended modification.
36. Screen transfected cell lines for the desired mutation:
    - For NHEJ, screen for small insertions or deletions (indels) using a mismatch-sensitive nuclease such as the T7 endonuclease or the CEL-1 surveyor assays.
    - For dual gRNA, PCR-amplify the target region and analyze on a 2.5% agarose gel to detect the deletion between the two gRNAs as a smaller band.

      Inversions are also possible but at a much lower frequency and can be screened for by a primer designed between the two gRNAs.
    - For HDR, use Sanger sequencing or genotype-specific primers to screen for the targeted mutation.

      Alternatively, if the alteration disrupts or creates a known restriction enzyme site that can distinguish the two genotypes, use that site to digest and identify targeted clones with the desired mutation.
37. Screen for both alleles by TOPO cloning the PCR products to sequence each allele individually to determine its genotype.

TROUBLESHOOTING

Problem (Step 15): There is a high rate of cell death after electroporation.
Solution: It is possible that the PSCs were too confluent before Accutase digestion. Ideally, electroporate when the cells are still in their exponential growth phase (i.e., 60%–80% confluency). Poor survival can also occur if the DNA exceeds 10% of the total volume of nucleofection solution. If this is the case, concentrate the plasmid further using ethanol precipitation or column elution to ensure a concentration >2 mg/mL.

Problem (Step 32): Mosaics composed of a mixture of wild-type cells with one or more cells with the desired modification are observed as when screening colonies picked after genome editing.
Solution: Repeat clonal isolation (Steps 27–31) and screen the new subclones for further enrichment of cells with the desired modification. Usually one additional round of colony picking will yield a clonal population but the process may be repeated as necessary to isolate a colony from a single cell.

RELATED INFORMATION

A more detailed discussion on the background and uses of these techniques is available in Introduction: Genome Editing in Human Pluripotent Stem Cells (Smith et al. 2016).

RECIPE

Phosphate-Buffered Saline (PBS)

Reagent	Amount to add (for 1× solution)	Final concentration (1×)	Amount to add (for 10× stock)	Final concentration (10×)
NaCl	8 g	137 mM	80 g	1.37 M
KCl	0.2 g	2.7 mM	2 g	27 mM
Na_2HPO_4	1.44 g	10 mM	14.4 g	100 mM
KH_2PO_4	0.24 g	1.8 mM	2.4 g	18 mM
If necessary, PBS may be supplemented with the following:				
$CaCl_2 \cdot 2H_2O$	0.133 g	1 mM	1.33 g	10 mM
$MgCl_2 \cdot 6H_2O$	0.10 g	0.5 mM	1.0 g	5 mM

PBS can be made as a 1× solution or as a 10× stock. To prepare 1 L of either 1× or 10× PBS, dissolve the reagents listed above in 800 mL of H_2O. Adjust the pH to 7.4 (or 7.2, if required) with HCl, and then add H_2O to 1 L. Dispense the solution into aliquots and sterilize them by autoclaving for 20 min at 15 psi (1.05 kg/cm^2) on liquid cycle or by filter sterilization. Store PBS at room temperature.

REFERENCES

Mali P, Yang L, Esvelt KM, Aach J, Guell M, DiCarlo JE, Norville JE, Church GM. 2013. RNA-guided human genome engineering via Cas9. *Science* **339:** 823–826.

Smith C, Ye Z, Cheng L. 2016. Genome editing in human pluripotent stem cells. *Cold Spring Harb Protoc* doi: 10.1101/pdb.top086819.

Yang L, Yang JL, Byrne S, Pan J, Church GM. 2014. CRISPR/Cas9-directed genome editing of cultured cells. *Curr Protoc Mol Biol* **107:** 31.1.1–31.1.17.

Cite this protocol as *Cold Spring Harb Protoc*; doi:10.1101/pdb.prot090217

CHAPTER 12

An Introduction to CRISPR Technology for Genome Activation and Repression in Mammalian Cells

Dan Du[1] and Lei S. Qi[1,2,3,4]

[1]*Department of Bioengineering, Stanford University, Stanford, California 94305;* [2]*Department of Chemical and Systems Biology, Stanford University, Stanford, California 94305;* [3]*ChEM-H; Stanford University, Stanford, California 94305*

CRISPR interference/activation (CRISPRi/a) technology provides a simple and efficient approach for targeted repression or activation of gene expression in the mammalian genome. It is highly flexible and programmable, using an RNA-guided nuclease-deficient Cas9 (dCas9) protein fused with transcriptional regulators for targeting specific genes to effect their regulation. Multiple studies have shown how this method is an effective way to achieve efficient and specific transcriptional repression or activation of single or multiple genes. Sustained transcriptional modulation can be obtained by stable expression of CRISPR components, which enables directed reprogramming of cell fate. Here, we introduce the basics of CRISPRi/a technology for genome repression or activation.

BACKGROUND

Targeted genome activation or repression is an important approach for engineering complex cellular functions, reprogramming cell fate and for disease modeling. In the past, RNA interference (RNAi) has been used as a major method for silencing the expression of genes in mammalian cells. RNAi uses base-pairing between small RNAs and mRNAs for triggering degradation of target transcripts (Chang et al. 2006). Protein-based tools such as zinc-fingers and transcription-activator-like effectors (TALEs) also provide customizable tools for site-specific perturbation of gene expression when fused to transcriptional activators or repressors (Kabadi and Gersbach 2014). However, these techniques have limited usefulness when compared with the emerging CRISPR technology owing to either high off-target effects (in the case of RNAi) or the difficulty experienced in their construction and delivery into cells (in the case of zinc fingers and TALEs). In contrast, the CRISPR technology offers a more efficient, robust, multiplexable, and designable approach for genome-wide activation or repression (Gilbert et al. 2013, 2014; Qi et al. 2013; Tanenbaum et al. 2014; Zalatan et al. 2015).

The CRISPR system for gene activation or repression has been repurposed from natural type II CRISPR systems in bacteria. We have named this CRISPR technology for gene regulation as "CRISPR interference" (CRISPRi for repression) or "CRISPR activation" (CRISPRa for activation). Both CRISPRi and CRISPRa use a catalytically inactive form of the Cas9 protein, termed dCas9, fused with transcriptional repressors and activators, respectively. Targeting of dCas9 to the genome is dictated by a single guide RNA (sgRNA) containing a designed 20-nucleotide sequence complementary to the DNA target, which is adjacent to a short DNA motif, termed the protospacer-adjacent motif (PAM; Fig. 1). Different homologs of Cas9 recognize different PAM sequences. For example, *Strepto-*

[4]Correspondence: stanley.qi@stanford.edu

Copyright © Cold Spring Harbor Laboratory Press; all rights reserved
Cite this introduction as *Cold Spring Harb Protoc*; doi:10.1101/pdb.top086835

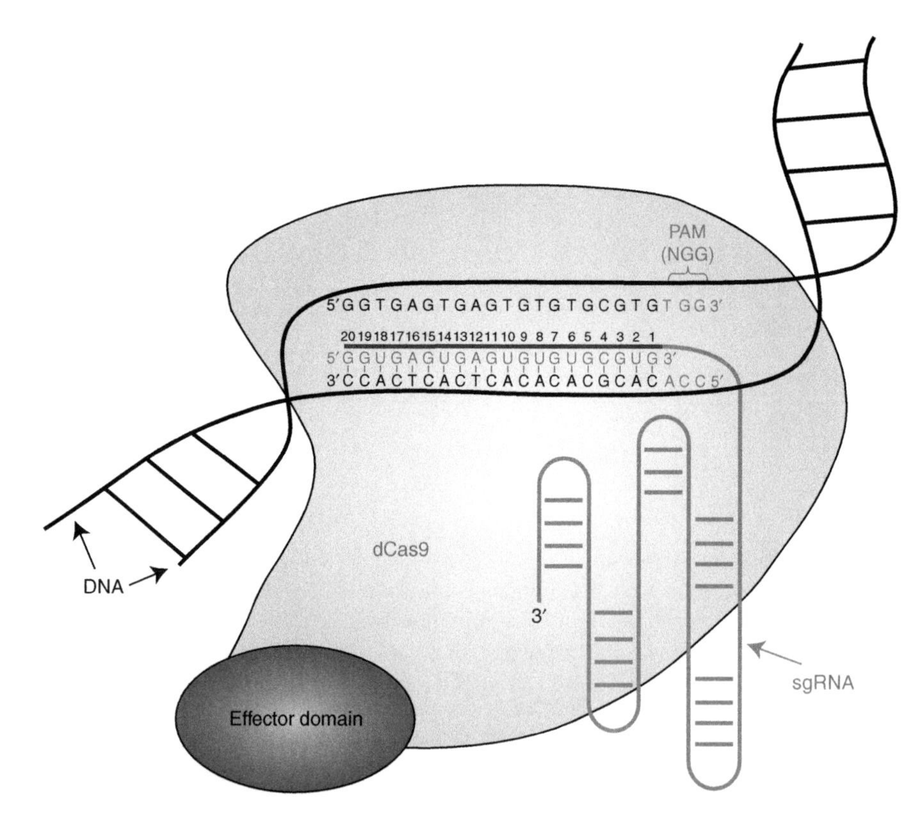

FIGURE 1. Illustration of sgRNA-guided DNA binding of dCas9. The diagram shows that a single guide RNA (sgRNA) can direct *S. pyogenes* nucleolytically deactivated Cas9 (dCas9, with D10A and H840A mutations) to a specific locus in the genome. The 20-bp guide sequence at the 5′ end of the sgRNA specifies the target sequence according to the rules of Watson–Crick base-pairing. On the target genomic DNA, the adjacent 5′-NGG PAM motif after the 20-bp target sequence is required to efficiently direct the dCas9–sgRNA complex to the genomic DNA. The numbers on top of the base-pairing targeting sequence indicate the nucleotide position relative to that of the adjacent PAM sequence. PAM, protospacer-adjacent motif; sgRNA, single guide RNA.

coccus pyogenes Cas9 recognizes NGG or NAG, whereas *Neisseria meningitides* Cas9 recognizes NNNNGATT (Hou et al. 2013). The dCas9 fusion protein complexed with a sequence-specific sgRNA binds to the target DNA and is engineered such that it localizes a repressive or activating effector domain to turn down or turn on transcription of the target genes. Armed with CRISPRi- and CRISPRa-based approaches, investigators now have a platform for systematically interrogating gene expression.

CRISPRi AND CRISPRa FOR TRANSCRIPTIONAL MODULATION

The CRISPR–dCas9 system provides a flexible and modular platform for recruiting different functional effector domains to essentially anywhere in the genome. This is mostly achieved by fusing dCas9 to different effector domains. Sometimes, recruitment of multiple effector domains is required for optimal genome-regulatory effectiveness.

CRISPRi

It has been established that the dCas9 protein alone can efficiently silence gene expression in bacteria with a properly designed sgRNA (Qi et al. 2013). However, only moderate repression was observed using dCas9 alone in mammalian cells. Strong gene silencing was seen when fusing the dCas9 protein to the repressive KRAB (Kruppel-associated box) domain of Kox1 (Gilbert et al. 2013). Target genes can be stably silenced when cells stably express dCas9–KRAB and an sgRNA (Gilbert et al. 2013). When multiple sgRNAs are coexpressed, multiple genes can be silenced simultaneously (see below).

 Cite this introduction as *Cold Spring Harb Protoc*; doi:10.1101/pdb.top086835

Effective targeting sites of CRISPRi include enhancers, proximal promoters, and the coding region downstream from the transcription start site (TSS) of a gene (Gilbert et al. 2013; Kearns et al. 2014).

CRISPRa

It has been shown that the dCas9 fusion with a transcription activator VP64 can activate a reporter gene relatively effectively (Cheng et al. 2013; Gilbert et al. 2013; Maeder et al. 2013; Mali et al. 2013; Perez-Pinera et al. 2013; Chakraborty et al. 2014; Kearns et al. 2014; Chavez et al. 2015). However, direct fusion of dCas9 to VP64 results in only very mild activation of endogenous target genes. For better activation of transcription, several systems have been developed. For example, fusing VP64 to both the amino and carboxyl terminus of dCas9 or fusing 10 copies of VP16 to dCas9 in each case enhanced activation (Cheng et al. 2013; Chakraborty et al. 2014). Chavez and colleagues generated a VP64–p65–Rta tripartite activator with dCas9 and showed that this construct enabled efficient endogenous gene activation (Chavez et al. 2015). All of these methods increased transcription activation of the target genes compared with that of the dCas9–VP64 system.

Additionally, a series of systems for indirect fusions of effector domains to CRISPR–Cas9 have been developed. For example, Konermann and colleagues appended two MS2 bacteriophage coat-protein-binding RNA motifs to two sgRNA stem loops. They coexpressed an MS2-activator (MS2–p65–HSF1) fusion protein together with dCas9–VP64 and modified sgRNA and observed more efficient transcription activation compared with dCas9–VP64 (Konermann et al. 2015). Tanenbaum and colleagues have developed a "SunTag" scaffold protein that can specifically recruit multiple copies of single-chain variable fragment (scFv), an artificial antibody fusion protein. When fusing scFv to a VP64 activator, multiple transcriptional activators can be recruited by dCas9–SunTag to the target DNA for very strong activation of endogenous genes (Gilbert et al. 2014; Tanenbaum et al. 2014).

Notably, CRISPRi and CRISPRa have low off-target effects. Using RNA-seq to assay the transcriptome, it has been shown that CRISPRi and CRISPRa can specifically modulate gene expression while inducing minimal off-target effects (Cheng et al. 2013; Gilbert et al. 2013, 2014; Perez-Pinera et al. 2013; Konermann et al. 2015).

We provide a working protocol for designing, cloning, and using sgRNAs for effective gene activation and repression in mammalian cells in Protocol 1: CRISPR Technology for Genome Activation and Repression in Mammalian Cells (Du and Qi 2016).

Modulation of Multiple Genes Using CRISPRi and CRISPRa

Multiple genes can be simultaneously activated or repressed by co-delivering multiple cognate sgRNAs, thus providing a powerful platform for analyzing the interaction of multiple genes (Cheng et al. 2013; Gilbert et al. 2013; Qi et al. 2013; Chavez et al. 2015; Konermann et al. 2015). To simultaneously activate and repress multiple genes in the same cell, scaffold RNAs (scRNAs) have been engineered by fusing sgRNAs to orthogonal protein-binding bacteriophage RNAs such as MS2, PP7, and Com (Zalatan et al. 2015). It has been shown that co-delivery of MCP–VP64 and COM–KRAB with a dCas9 protein allows simultaneous activation of *CXCR4* (a chemokine receptor) and repression of *B4GAL4NT1* (encoding β-1,4-*N*-acetyl-galactosaminyl transferase) in the same cell (Zalatan et al. 2015). Thus, engineered scRNAs provide a versatile platform for multigene modulation for recruiting diverse effectors to different genomic loci.

Repression and Activation of Noncoding RNA Genes

In addition to protein-coding genes, CRISPRi can be harnessed to repress transcription of long noncoding RNAs (lncRNAs). For example, strong knockdown (>80%) of five tested lncRNAs (*H19*, *MALAT1*, *NEAT1*, *TERC*, and *XIST*) has been observed in human myelogenous leukemia K562 cells (Gilbert et al. 2014). These results showed that CRISPRi was able to repress lncRNA expression effectively, enabling further functional analysis of these noncoding genes. Konermann and colleagues have also shown that CRISPRa can activate long intergenic noncoding RNAs (lincRNAs) such as *TINCR*, *PCAT*, and *HOTTIP* (Konermann et al. 2015).

Application of CRISPRi and CRISPRa for Reprogramming of Cell Fate

Expression of exogenous transcription factors has been used as a major approach for directed cell reprogramming (Ladewig et al. 2013). Now, CRISPR-based gene regulation provides a novel approach. For example, Kearns and colleagues showed that CRISPRi could modulate the differentiation of human pluripotent stem cells by using an sgRNA to repress the *OCT4* gene (Kearns et al. 2014). Furthermore, Chakraborty and colleagues showed that CRISPRa could induce the transdifferentiation of mouse embryonic fibroblasts into skeletal myocytes by activating transcription of the endogenous *Myod1* gene (Chakraborty et al. 2014). Finally, when paired with sgRNAs targeting the *OCT4* promoter, dCas9–VP192 has been used to replace the requirement for exogenous *OCT4* overexpression in a methodology for reprogramming human-induced pluripotent stem cells (Balboa et al. 2015).

In summary, CRISPRi and CRISPRa offer powerful approaches for repression and activation of endogenous genes, which is useful for studying gene functions, rewiring genetic networks, and reprogramming cell fates.

ACKNOWLEDGMENTS

We thank Antonia Dominguez, Marie La Russa, and Yanxia Liu for critical comments on the manuscript. The authors acknowledge support from the California Institute for Quantitative Biomedical Research (QB3), National Institutes of Health Office of The Director (OD), and National Institute of Dental and Craniofacial Research (NIDCR). This work was supported by National Institutes of Health Director's Early Independence Award (grant OD017887 L.S.Q.).

REFERENCES

Balboa D, Weltner J, Eurola S, Trokovic R, Wartiovaara K, Otonkoski T. 2015. Conditionally stabilized dCas9 activator for controlling gene expression in human cell reprogramming and differentiation. *Stem Cell Reports* **5:** 448–459.

Chakraborty S, Ji H, Kabadi AM, Gersbach CA, Christoforou N, Leong KW. 2014. A CRISPR/Cas9-based system for reprogramming cell lineage specification. *Stem Cell Reports* **3:** 940–947.

Chang K, Elledge SJ, Hannon GJ. 2006. Lessons from nature: MicroRNA-based shRNA libraries. *Nat Methods* **3:** 707–714.

Chavez A, Scheiman J, Vora S, Pruitt BW, Tuttle M, PR Iyer E, Lin S, Kiani S, Guzman CD, Wiegand DJ, et al. 2015. Highly efficient Cas9-mediated transcriptional programming. *Nat Methods* **12:** 326–328.

Cheng AW, Wang H, Yang H, Shi L, Katz Y, Theunissen TW, Rangarajan S, Shivalila CS, Dadon DB, Jaenisch R. 2013. Multiplexed activation of endogenous genes by CRISPR-on, an RNA-guided transcriptional activator system. *Cell Res* **23:** 1163–1171.

Du D, Qi LS. 2016. CRISPR technology for genome activation and repression in mammalian cells. *Cold Spring Harb Protoc* doi: 10.1101/pdb.prot090175.

Gilbert LA, Larson MH, Morsut L, Liu Z, Brar GA, Torres SE, Stern-Ginossar N, Brandman O, Whitehead EH, Doudna JA, et al. 2013. CRISPR-mediated modular RNA-guided regulation of transcription in eukaryotes. *Cell* **154:** 442–451.

Gilbert LA, Horlbeck MA, Adamson B, Villalta JE, Chen Y, Whitehead EH, Guimaraes C, Panning B, Ploegh HL, Bassik MC, et al. 2014. Genome-scale CRISPR-mediated control of gene repression and activation. *Cell* **159:** 647–661.

Hou Z, Zhang Y, Propson NE, Howden SE, Chu LF, Sontheimer EJ, Thomson JA. 2013. Efficient genome engineering in human pluripotent stem cells using Cas9 from *Neisseria meningitidis*. *Proc Natl Acad Sci* **110:** 15644–15649.

Kabadi AM, Gersbach CA. 2014. Engineering synthetic TALE and CRISPR/Cas9 transcription factors for regulating gene expression. *Methods* **69:** 188–197.

Kearns NA, Genga RM, Enuameh MS, Garber M, Wolfe SA, Maehr R. 2014. Cas9 effector-mediated regulation of transcription and differentiation in human pluripotent stem cells. *Development* **141:** 219–223.

Konermann S, Brigham MD, Trevino AE, Joung J, Abudayyeh OO, Barcena C, Hsu PD, Habib N, Gootenberg JS, Nishimasu H, et al. 2015. Genome-scale transcriptional activation by an engineered CRISPR-Cas9 complex. *Nature* **517:** 583–588.

Ladewig J, Koch P, Brustle O. 2013. Leveling Waddington: The emergence of direct programming and the loss of cell fate hierarchies. *Nat Rev Mol Cell Biol* **14:** 225–236.

Maeder ML, Linder SJ, Cascio VM, Fu Y, Ho QH, Joung JK. 2013. CRISPR RNA-guided activation of endogenous human genes. *Nat Methods* **10:** 977–979.

Mali P, Aach J, Stranges PB, Esvelt KM, Moosburner M, Kosuri S, Yang L, Church GM. 2013. CAS9 transcriptional activators for target specificity screening and paired nickases for cooperative genome engineering. *Nat Biotechnol* **31:** 833–838.

Perez-Pinera P, Kocak DD, Vockley CM, Adler AF, Kabadi AM, Polstein LR, Thakore PI, Glass KA, Ousterout DG, Leong KW, et al. 2013. RNA-guided gene activation by CRISPR-Cas9-based transcription factors. *Nat Methods* **10:** 973–976.

Qi LS, Larson MH, Gilbert LA, Doudna JA, Weissman JS, Arkin AP, Lim WA. 2013. Repurposing CRISPR as an RNA-guided platform for sequence-specific control of gene expression. *Cell* **152:** 1173–1183.

Tanenbaum ME, Gilbert LA, Qi LS, Weissman JS, Vale RD. 2014. A protein-tagging system for signal amplification in gene expression and fluorescence imaging. *Cell* **159:** 635–646.

Zalatan JG, Lee ME, Almeida R, Gilbert LA, Whitehead EH, La Russa M, Tsai JC, Weissman JS, Dueber JE, Qi LS, et al. 2015. Engineering complex synthetic transcriptional programs with CRISPR RNA scaffolds. *Cell* **160:** 339–350.

Cite this introduction as *Cold Spring Harb Protoc*; doi:10.1101/pdb.top086835

Protocol 1

CRISPR Technology for Genome Activation and Repression in Mammalian Cells

Dan Du[1] and Lei S. Qi[1,2,3,4]

[1]*Department of Bioengineering, Stanford University, Stanford, California 94305;* [2]*Department of Chemical and Systems Biology, Stanford University, Stanford, California 94305;* [3]*ChEM-H; Stanford University, Stanford, California 94305*

Targeted modulation of transcription is necessary for understanding complex gene networks and has great potential for medical and industrial applications. CRISPR is emerging as a powerful system for targeted genome activation and repression, in addition to its use in genome editing. This protocol describes how to design, construct, and experimentally validate the function of sequence-specific single guide RNAs (sgRNAs) for sequence-specific repression (CRISPRi) or activation (CRISPRa) of transcription in mammalian cells. In this technology, the CRISPR-associated protein Cas9 is catalytically deactivated (dCas9) to provide a general platform for RNA-guided DNA targeting of any locus in the genome. Fusion of dCas9 to effector domains with distinct regulatory functions enables stable and efficient transcriptional repression or activation in mammalian cells. Delivery of multiple sgRNAs further enables activation or repression of multiple genes. By using scaffold RNAs (scRNAs), different effectors can be recruited to different genes for simultaneous activation of some and repression of others. The CRISPRi and CRISPRa methods provide powerful tools for sequence-specific control of gene expression on a genome-wide scale to aid understanding gene functions and for engineering genetic regulatory systems.

MATERIALS

It is essential that you consult the appropriate Material Safety Data Sheets and your institution's Environmental Health and Safety Office for proper handling of equipment and hazardous materials used in this protocol.

RECIPES: Please see the end of this protocol for recipes indicated by <R>. Additional recipes can be found online at http://cshprotocols.cshlp.org/site/recipes.

Reagents

Chemically competent *Escherichia coli* cells (e.g., One Shot TOP10 Cells from Life Technologies)

dCas9 expression vector(s) appropriate for experiment

- CRISPR activation (CRISPRa) dCas9–SunTag expression vectors

  Two constructs are required: a lentiviral vector containing an SV40-promoter-driven dCas9 fusion between dCas9, 2X nuclear localization signal (NLS), 10X GCN4, and a P2A-tagBFP (Addgene 60903) and a lentiviral vector containing an SV40-promoter-driven fusion protein between the single chain variable fragment (scFv) for GCN4, a superfolder (sf) GFP, VP64, and 1X NLS (Addgene 60904).

- CRISPR interference (CRISPRi) dCas9–KRAB expression vector

  This comprises a lentiviral vector containing a spleen focus-forming virus SFFV-promoter-driven dCas9 fused to 2X NLS, a tagBFP and a KRAB domain (Addgene 46911).

[4]Correspondence: stanley.qi@stanford.edu

Copyright © Cold Spring Harbor Laboratory Press; all rights reserved

Cite this protocol as *Cold Spring Harb Protoc*; doi:10.1101/pdb.prot090175

dNTPs (10 mM)
Double-distilled water (ddH_2O), sterile and nuclease-free
Dulbecco's modified Eagle's medium (DMEM), high-glucose (Life Technologies 11965-092)
Fetal bovine serum (FBS)
Gel electrophoresis reagents

- Agarose gels (1%, w/v)
- DNA ladder
- Ethidium bromide
- Tris-acetate-EDTA (TAE) buffer (50×) <R>

HEK293T cells (ATCC CRL-11268)

HEK293T cells (or other cells derived from HEK293T cells) are required for lentivirus production in Steps 26–31. In addition, they are used here as an example of target cells in Steps 32–36. Other target cells may be used as appropriate for the experiment.

In-Fusion HD Cloning Kit (Clontech 011614)
iQ SYBR Green Supermix (Bio-Rad 170-8880)
iScript cDNA Synthesis Kit (Bio-Rad 170-8890)
Lentiviral packaging plasmids pCMV-dR8.91 and pMD2.G (Addgene 12259)

At the time of this writing, pCMV-dR8.91 is not avialable from Addgene. Alternatively, a lower version of the plasmid is available from Addgene (pCMV-dR8.2; Addgene 8455). Its use will not affect this protocol.

Lysogeny broth (LB) with carbenicillin (liquid medium and agar plates) <R>
Mirus TransIT-LT1 Transfection Reagent (Mirus MIR 2300)
Opti-MEM Reduced-Serum Medium (Life Technologies 31985-062)
Penicillin-Streptomycin (100×), presterilized (Life Technologies 15070-063)
Phusion High-Fidelity Polymerase and 5× Phusion HF Buffer (New England BioLabs M0536L)
Polybrene (optional; see Step 33)
Primers

- PCR primers, one of which (sgRNA-F) contains the gene-specific sgRNA target sequence
  Forward primer (sgRNA-F): 5′-CCCTTGGAGAACCACCTTGTTGG$N_{(19)}$GTTTAAGAGCTATGCTGGAAACAGCA-3′
  Reverse primer (sgRNA-R): 5′-GATCCTAGTACTCGAGAAAAAAAGCACCGACTCGGTGCCAC-3′

For sgRNA target sequence selection, see Steps 1–3.

- Sequencing primer: 5′-GAGGCTTAATGTGCGATAAAAGA-3′

  This primer binds to the mouse U6 promoter and is used to confirm the generation of sgRNA expression constructs in Step 21.

- Target gene-specific primers for qRT-PCR (see Step 40)

QIAGEN Plasmid Midi Kit (QIAGEN 12143)

It is important to use an endotoxin-free midiprep kit when purifying plasmid DNA for better transfection efficiency into mammalian cells.

QIAprep Spin Miniprep Kit (QIAGEN 27106)
QIAquick Gel Extraction Kit (QIAGEN 28706)
QIAquick PCR Purification Kit (QIAGEN 28106)
Restriction enzymes BstXI, XhoI, and DpnI
RNeasy Plus Mini Kit (QIAGEN 74134)
Single guide RNA (sgRNA) expression vector

This comprises a lentiviral vector containing the mouse U6 promoter driving sgRNA expression (Addgene 51024). It also contains an expression cassette consisting of a cytomegalovirus (CMV) promoter, a puromycin-resistance gene cassette, and an mCherry gene for selection purposes.

Trypsin-EDTA (0.05%) (e.g., Life Technologies 25300-054)

Cite this protocol as *Cold Spring Harb Protoc*; doi:10.1101/pdb.prot090175

Equipment

Access to sequencing facility (see Step 21)
BD FACSAria II Cell Sorter (BD Biosciences) equipped with lasers and filters for detecting mCherry, EGFP, and tagBFP
CFX96 Real-Time PCR Detection System (Bio-Rad 185-5195)
CO_2 incubator at 37°C and 5% CO_2 for mammalian cell culture
Computer with Internet-connected web browser
Conical tubes
Digital gel-imaging system
Erlenmeyer flasks (250 mL)
Gel electrophoresis system
Glass tubes (25-mm)
Incubators at 37°C for growing bacteria (one standard and one capable of shaking at 200 rpm)
Microcentrifuge
Microcentrifuge tubes
Microplate for qRT-PCR
NanoDrop 8000 UV-Vis Spectrophotometer (Thermo Scientific)
PCR tubes (0.2-mL)
Six-well tissue-culture plates
Syringe filter (0.45-µm), sterile
Syringe, sterile
Thermocycler

METHOD

We have implemented a computational tool, termed CRISPR-ERA ("editing, repression, and activation") for automated design of sgRNAs for given mammalian organisms, such as mouse, rat, and human (Liu et al. 2015). The CRISPR-ERA algorithm aligns the designed sgRNA to the whole genome and reports potential off-target sites as defined by possession of fewer than three mismatches. The tool is freely available at http://CRISPR-ERA.stanford.edu. If using CRISPR-ERA, skip Steps 1–5.

Selection of sgRNA Targets in the Genome

1. Determine the DNA sequence of the target gene using an available genome database—for example, the UCSC genome browser (Kent et al. 2002).
2. Obtain annotation information of the target gene, including the location of the transcription start site (TSS).
3. Search for patterns of $GN_{(19)}NGG$ around the TSS, wherein $GN_{(19)}$ is the binding site of the sgRNA and NGG is the protospacer adjacent motif (PAM), which is required for efficient DNA binding of *Streptococcus pyogenes* Cas9.

   Our sgRNA expression construct uses a mouse U6 promoter, which requires a G at the very 5′ end for effective transcription. Therefore, we search for $GN_{(19)}$ as the binding site of the sgRNA. If another promoter is used, it is likely that the first nucleotide will be different.

   The recommended window of the target DNA is −50 to +300 bp relative to the TSS for CRISPR interference (CRISPRi) for gene repression, or −400 to −50 bp for CRISPR activation (CRISPRa). Usually, multiple sgRNA-binding sites within the target window of the gene need to be tested to define the most efficient targeting site for repression or activation.

   Many mammalian genes possess transcript isoforms with different TSSs. In this case, different sgRNAs need to be designed for each transcript. Currently, there is no direct evidence that the activities of CRISPRi and CRISPRa are sensitive to the DNA strand or GC content (Gilbert et al. 2014).

Design of sgRNA Sequences

4. Ensure that the base-pairing sequence on the sgRNA is the reverse complement of the $GN_{(19)}$ sequence identified in Step 3.
5. Analyze the specificity of the target sequence in the genome use the basic local alignment search tool (BLAST; http://blast.ncbi.nlm.nih.gov) (Bhagwat et al. 2012).

   The BLAST algorithm enables the specificity of sgRNA targeting in the genome to be analyzed when not using the CRISPR-ERA tool.
6. Generate the full-length sgRNA by appending $GN_{(19)}$ 3′ to the rest of the optimized sgRNA sequence (5′-$GN_{(19)}$GUUUAAGAGCUAUGCUGGAAACAGCATAGCAAGUUUAAAUAAGG CUAGUCCGUUAUCAACUUGAAAAAGUGGCACCGAGUCGGUGCUUUUUUU-3′) (Chen et al. 2013).
7. Confirm that the $GN_{(19)}$ target sequence does not contain any transcription termination sequence for the U6 promoter (Paul et al. 2002).

Preparation of sgRNA Expression Constructs

8. Generate the sgRNA backbone by digesting the empty sgRNA expression vector with restriction enzymes BstXI and XhoI for 4–16 h at 37°C according to the manufacturer's instructions.
9. Separate the digested sgRNA backbone products by electrophoresis through a 1% (w/v) agarose gel in 1× TAE buffer. Stain the gel with ethidium bromide, and visualize the bands using a digital gel imaging system. Compare the bands to those of a proper DNA ladder, and confirm that the band representing the sgRNA backbone DNA is ~8 kb.
10. Gel-purify the sgRNA backbone DNA using a QIAquick Gel Extraction Kit according to the manufacturer's instructions. Store the DNA at −20°C until use in Step 15.
11. Perform PCR as follows, using primers that contain the 20-nt target sequences identified in Steps 1–3.

    i. Assemble the following reaction (volumes shown are for one reaction) in a 0.2-mL PCR tube on ice.

0.5 µL	Empty sgRNA expression vector (undigested) as template (100 ng/µL)
2.5 µL	Forward primer (sgRNA-F) (10 µM)
2.5 µL	Reverse primer (sgRNA-R) (10 µM)
2 µL	dNTPs (10 mM)
0.5 µL	Phusion High-Fidelity Polymerase (2 U/µL)
10 µL	Phusion HF Buffer (5×)
32 µL	Nuclease-free water
50 µL	Total volume

    ii. Perform PCR with the following cycling conditions:

1 cycle	98°C	30 sec
25 cycles	98°C	10 sec
	62°C	30 sec
	72°C	10 sec
1 cycle	72°C	5 min
1 cycle	4°C	Forever

12. Confirm that the PCRs successfully amplified a ~150-bp DNA product by separating 5 µL of the PCR products on a 1% agarose gel as in Step 9.

 Cite this protocol as *Cold Spring Harb Protoc*; doi:10.1101/pdb.prot090175

13. Add 1 µL of DpnI (20 U/µL) into each PCR and then incubate for 1 h at 37°C.

    Treatment of DpnI will digest the PCR templates.

14. Purify the PCRs using a QIAquick PCR Purification Kit by following the manufacturer's instructions. Store the DNA at −20°C until use in Step 15.
15. Measure the concentrations of the purified sgRNA backbone DNA (from Step 10) and PCR fragments (from Step 14) using a NanoDrop UV-Vis 8000 Spectrophotometer.
16. Ligate the PCR fragments to the sgRNA backbone DNA using an In-Fusion HD Cloning Kit.

    i. Assemble the cloning reaction.

1 µL	In-Fusion HD Enzyme Premix (5×)
50 ng	Linearized sgRNA backbone DNA (from Step 10)
25 ng	Purified PCR fragments (from Step 14)
x µL	ddH_2O
5 µL	Total volume

    ii. Incubate the reaction for 15 min at 50°C using a thermocycler.

    iii. Place on ice for 5 min. Store at −20°C until use in Step 17.

17. Transform chemically competent *E. coli* cells with the products of the ligation reactions. Follow the manufacturer's instructions for the *E. coli* cells. Spread transformed *E. coli* cells onto LB agar plates supplemented with 100 µg/mL carbenicillin. Incubate the plates overnight in a 37°C incubator.
18. Transfer single colonies into 25-mm glass tubes containing 5 mL of LB medium supplemented with 100 µg/mL carbenicillin. For each colony, use a sterile pipette tip to touch the colony, and then swirl the tip in the LB medium to dissolve the colony. Incubate overnight in a 37°C shaking incubator, swirling at 200 rpm.
19. Transfer 0.5 mL of bacterial culture into a 250-mL Erlenmeyer flask containing 50 mL of LB medium with 100 µg/mL carbenicillin. Incubate overnight in a 37°C shaking incubator, swirling at 200 rpm.
20. Extract the plasmid DNA from the remaining 4.5 mL of bacterial culture using a QIAprep Spin Miniprep Kit according to the manufacturer's instructions.
21. Send the extracted plasmid DNA for sequencing with the sequencing primer.
22. After the plasmid is verified by sequencing, extract DNA from the 50-mL bacterial culture using a QIAGEN Plasmid Midi Kit according to the manufacturer's instructions. Store the DNA at −20°C until use in Step 27.

Preparation of dCas9 Expression Vectors

23. Transform chemically competent *E. coli* cells with the dCas9 expression vectors appropriate for the experiment (CRISPRi or CRISPRa). Spread transformed *E. coli* cells onto LB agar plates supplemented with 100 µg/mL carbenicillin. Incubate the plates overnight in a 37°C incubator.
24. Transfer a single colony into 50 mL of LB medium supplemented with 100 µg/mL carbenicillin. Incubate overnight in a 37°C shaking incubator, swirling at 200 rpm.
25. Extract DNA using a QIAGEN Plasmid Midi Kit according to the manufacturer's instructions. Store the DNA at −20°C until use in Step 27.

Packaging of dCas9 and sgRNA Expression Constructs into Lentiviral Particles

If more lentiviruses are required, scale up the cell numbers, DNA amounts, and transfection reagent volumes used here.

26. On the day before transfection, seed a six-well tissue-culture plate with 2–3 × 10^5 HEK293T cells in 2 mL of high-glucose DMEM containing 10% (v/v) FBS per well. Incubate overnight at 37°C and 5% CO_2.

    HEK293T cells can be maintained in regular high-glucose DMEM medium supplemented with 10% (v/v) FBS, 100 U/mL streptomycin, and 100 µg/mL penicillin and regularly passaged using 0.05% (w/v) trypsin–EDTA. However, antibiotic-free DMEM is required during transfection and virus collection to achieve better efficiency.

27. Twenty-four hours after plating the cells, prepare the transfection complexes as follows.

    i. Combine the following DNA samples.

1.32 µg	pCMV-dR8.91 (lentiviral packaging plasmid)
165 ng	pMD2.G (lentiviral packaging plasmid)
1.51 µg	dCas9 or sgRNA expression construct

    Nontargeting sgRNA vector or dCas9 without fusion vector can be used in parallel as a negative control.

    ii. Add this 3-µg DNA mixture into 250 µL of Opti-MEM Reduced-Serum Medium in a microcentrifuge tube. Mix well by pipetting up and down.

    iii. Add 7.5 µL of Mirus TransIT-LT1 Transfection Reagent into the same tube. Mix well by pipetting up and down.

    iv. Allow transfection complexes to form for 30 min at room temperature.

28. Remove 250 µL of medium from each well in the six-well plate.

29. Add the ~250-µL mixture from Step 27.iv into one well in the six-well plate. Mix well by rocking the plate gently back and forth. Incubate for 24 h at 37°C and 5% CO_2.

    Cells will begin producing viruses 24 h after transfection.

30. After the 24-h incubation, replace the transfection medium with 2.5 mL of fresh DMEM with 10% FBS.

    If the target cells to be infected have any additional medium requirements, replace the transfection medium with 2.5 mL of special growth medium for the target cells.

31. Use a sterile syringe to harvest the viral supernatant 24–48 h after medium replacement. Filter the medium through a 0.45-µm syringe filter into a conical tube to avoid transferring HEK293T cells.

    The total volume will be ~2 mL after filtering. Lentiviral particles can be stored for up to 1 wk at 4°C, or snap-frozen in liquid nitrogen and stored for several months at −80°C. However, we recommend using the lentiviruses immediately after collection.

Transduction of Target Cells with dCas9 and sgRNA Lentiviral Particles

In the following, the use of HEK293T cells is given as an example. For other types of cells, modify the procedure (e.g., cell number and growth medium) as appropriate.

32. Sixteen hours before transduction, seed a six-well tissue-culture plate with 1.5–2 × 10^5 HEK293T cells in 2 mL of high-glucose DMEM supplemented with 10% FBS per well. Incubate at 37°C and 5% CO_2.

33. Replace the medium with 1 mL of DMEM containing 10% FBS and 1 mL of filtered viral supernatant. Incubate overnight at 37°C and 5% CO_2.

    Depending on the virus titration, the viral supernatant can be diluted with growth medium for the target cells. Polybrene can be used to promote the infection efficiency with proper concentration; however, it is toxic for some types of cells, including HEK293T cells.

34. Replace the viral supernatant with 2 mL of fresh DMEM with 10% FBS, and incubate for 48 h at 37°C and 5% CO_2.

 Cite this protocol as *Cold Spring Harb Protoc*; doi:10.1101/pdb.prot090175

Cells usually will express dCas9 protein 48 h after addition of lentiviruses. However, for repression experiments, we suggest collecting cells at least 72 h after infection to minimize the interference by preexisting target gene mRNA. If necessary, split the cells when they reach 80%–90% confluence before sorting.

35. Use a BD FACSAria II Cell Sorter to collect the cells.

    - For the CRISPRi system, collect cells positive for both blue fluorescent protein (BFP) and mCherry.

        The BFP-positive cells should express dCas9 protein, and mCherry-positive cells should express sgRNA.

    - For the CRISPRa (dCas9–Suntag) system, collect cells that are positive for BFP, mCherry, and GFP.

        The GFP-positive cells should express scFv-sfGFP-VP64 fusion protein.

36. Incubate the collected cells at 37°C and 5% CO_2.

    After the cells are grown up, analyze the expression levels of target genes by qRT-PCR as described in Steps 37–40.

Quantification of the Effects of CRISPRi or CRISPRa on Gene Expression in Target Cells

37. Extract total RNA from the cells infected using an RNeasy Plus Mini Kit according to the manufacturer's instructions.

    Typically, 0.5–1 × 10^6 cells (50%–80% confluence of cells in one well in a six-well plate) are sufficient for total RNA extraction.

38. Measure the concentrations of the total RNA samples using a NanoDrop spectrophotometer.
39. Synthesize cDNA using an iScript cDNA Synthesis Kit.

    i. Set up the cDNA synthesis reaction.

4 µL	iScript reaction mix (5×)
1 µL	iScript reverse transcriptase
1 µg	Total RNA template
x µL	Nuclease-free water
20 µL	Total volume

    ii. Incubate the reaction as follows (e.g., using a thermocycler):

1 cycle	25°C	5 min
1 cycle	42°C	30 min
1 cycle	85°C	5 min
1 cycle	4°C	Forever

    iii. Store the DNA at −20°C until use in Step 40.i.

40. Analyze the cDNA levels of target genes using a standard qRT-PCR protocol.

    i. Set up the PCR in a microplate using the iQ SYBR Green Supermix according to the manufacturer's instructions.

10 µL	iQ SYBR Green Supermix (2×)
1.2 µL	Forward primer (target gene-specific; 5 µM)
1.2 µL	Reverse primer (target gene-specific; 5 µM)
0.25 µL	Template (cDNA from Step 39)
7.35 µL	Nuclease-free water
20 µL	Total volume

    The amount of template cDNA can be scaled up or down according to the expression levels of the target genes in the cells. Housekeeping genes—for example, GAPDH, encoding glyceraldehyde-3-phosphate dehydrogenase—should be used as references.

ii. Run the following real-time PCR profile in a CFX96 Real-Time PCR Detection System.

1 cycle	95°C	2–3 min
39 cycles	95°C	10–15 sec
	55°C–60°C	15–30 sec
	72°C	30 sec
Melt curve (optional)	55°C–95°C (in 0.5°C increments)	10–30 sec

iii. Analyze the qRT-PCR data by standard methods to obtain the relative transcriptional expression levels of the target genes regulated by CRISPRi/a.

We use the $2^{-\Delta\Delta Ct}$ method to obtain the relative mRNA expression level of the CRISPRi or CRISPRa sample vs. a control sample, where $\Delta\Delta Ct = \Delta Ct$(CRISPRi/a sample)–$\Delta Ct$(control sample, e.g., nontargeting sgRNA sample), and where ΔCt(sample) = Ct(any sample)–Ct(endogenous housekeeping gene).

DISCUSSION

To date, several tools have been developed to functionally interrogate gene expression. RNAi has been shown to disrupt gene expression by triggering the degradation of target mRNAs (Chang et al. 2006). However, the technique is somewhat limited in its application owing to off-target effects and through being restricted to cytosolic target mRNAs (Jackson et al. 2003; Adamson et al. 2012; Sigoillot et al. 2012). Protein-based tools are difficult to be designed, cloned, and delivered into target cells. The complex programming and limited targeting sites also restrict the application of zinc fingers and tools based on transcription-activator-like effectors (TALEs) for perturbing the expression of multiple genes. Loss-of-function approaches based on genome editing, such as CRISPR–Cas9, cause irreversible frameshift disruptions, cytotoxic double-stranded DNA breaks, and in-frame insertion–deletions (indels) arising from error-prone DNA repair. These could limit the ability of the CRISPR technique to completely abolish the function of genes and noncoding RNAs (Huang et al. 1996; Jackson 2002; Koike-Yusa et al. 2014; Shalem et al. 2014; Wang et al. 2014).

In contrast, RNA-guided DNA targeting of the dCas9 protein to a specific locus provides a programmable platform to modulate genome status while generating minimal off-target effects. Fusion of different effector domains to dCas9 enables transcriptional repression (CRISPRi) or activation (CRISPRa) of specific target genes. CRISPRi and CRISPRa enable inducible and reversible modulation of specific endogenous gene expression within an intact biological system. The modulation of the transcription of single or multiple genes can be specifically achieved by delivery of multiple sgRNAs (Gilbert et al. 2013, 2014; Qi et al. 2013; Tanenbaum et al. 2014; Zalatan et al. 2015). By using scRNA, transcriptional activation or repression of different target genes can be achieved simultaneously in the same cell (Zalatan et al. 2015). Recently, CRISPRa has been used to effectively activate expression of target genes in plants and flies (Lin et al. 2015; Lowder et al. 2015). Furthermore, Kleinstiver and colleagues have modified *S. pyogenes* Cas9 (spCas9) to recognize alternative PAM sequences (other than NGG) by using a selection-based approach in bacterial cells (Kleinstiver et al. 2015). This provides researchers an expanded targetable sequence space in the genome for using CRISPR–dCas9. Thus, owing to its simplicity and flexibility, CRISPRi or CRISPRa can facilitate genome-scale perturbation of gene expression (Gilbert et al. 2014; Konermann et al. 2015).

However, the detailed mechanism underlying how CRISPRi and CRISPRa components interact with local transcriptional machinery and epigenetic factors is not well established. We usually design three to five sgRNAs for each target transcript and choose the best one for functional analysis. The reason why some of the designed sgRNAs have no function and why the efficiency of different designed sgRNAs varies is not clear. Knowledge of the mechanism would assist the efficiency of designing functional sgRNAs. Moreover, the spCas9 protein, which is widely used for transcriptional modulation, is a large molecule that is difficult to clone and package with the necessary regulatory elements into a size-restricted virus, such as the adeno-associated virus (AAV) that has been generally

 Cite this protocol as *Cold Spring Harb Protoc*; doi:10.1101/pdb.prot090175

used for gene therapy. In contrast, the smaller ortholog *Staphylococcus aureus* Cas9 (saCas9) has been shown to edit the targets efficiently and to be compatible with the AAV system, which has also been engineered as a transcriptional activating system (SAM) (Nishimasu et al. 2015; Ran et al. 2015). Currently, the gene-regulatory tools based on the *S. aureus* dCas9 are being developed for more-efficient transcriptional repression and activation. Thus, in summary, CRISPRi and CRISPRa based on different species of Cas9 or its homologs provide a versatile platform to manipulate and interrogate gene expression systematically.

RECIPES

Lysogeny Broth (LB) with Carbenicillin

Reagent	Quantity
Agar (for plates only)	20 g
NaCl	10 g
Tryptone	10 g
Yeast extract	5 g

Prepare the above-listed ingredients in 1 L of deionized water. Adjust the pH to 7.0 with 5 N NaOH. Autoclave for 20 min at 15 psi (1.05 kg/cm^2). Cool to ~60°C and add carbenicillin (final concentration 100 µg/mL). Pour the medium into Petri dishes (~25 mL per 100-mm plate). Store the LB plates at 4°C; they will keep for at least 4 mo.

Tris-Acetate-EDTA (TAE) Buffer (50×)

Reagent	Final concentration (1×)
Tris base	40 mM
EDTA	2 mM
Acetic acid	20 mM

Adjust to pH 8.5 and dilute to 1× with Milli-Q H_2O before use.

REFERENCES

Adamson B, Smogorzewska A, Sigoillot FD, King RW, Elledge SJ. 2012. A genome-wide homologous recombination screen identifies the RNA-binding protein RBMX as a component of the DNA-damage response. *Nat Cell Biol* **14:** 318–328.

Bhagwat M, Young L, Robison R.R. 2012. Using BLAT to find sequence similarity in closely related genomes. *Curr Protoc Bioinformatics* **Chapter 10:** Unit 10.18.

Chang K, Elledge SJ, Hannon GJ. 2006. Lessons from Nature: MicroRNA-based shRNA libraries. *Nat Methods* **3:** 707–714.

Chen B, Gilbert LA, Cimini BA, Schnitzbauer J, Zhang W, Li GW, Park J, Blackburn EH, Weissman JS, Qi LS, et al. 2013. Dynamic imaging of genomic loci in living human cells by an optimized CRISPR/Cas system. *Cell* **155:** 1479–1491.

Gilbert LA, Larson MH, Morsut L, Liu Z, Brar GA, Torres SE, Stern-Ginossar N, Brandman O, Whitehead EH, Doudna JA, et al. 2013. CRISPR-mediated modular RNA-guided regulation of transcription in eukaryotes. *Cell* **154:** 442–451.

Gilbert LA, Horlbeck MA, Adamson B, Villalta JE, Chen Y, Whitehead EH, Guimaraes C, Panning B, Ploegh HL, Bassik MC, et al. 2014. Genome-scale CRISPR-mediated control of gene repression and activation. *Cell* **159:** 647–661.

Huang LC, Clarkin KC, Wahl GM. 1996. Sensitivity and selectivity of the DNA damage sensor responsible for activating p53-dependent G1 arrest. *Proc Natl Acad Sci* **93:** 4827–4832.

Jackson SP. 2002. Sensing and repairing DNA double-strand breaks. *Carcinogenesis* **23:** 687–696.

Jackson AL, Bartz SR, Schelter J, Kobayashi SV, Burchard J, Mao M, Li B, Cavet G, Linsley PS. 2003. Expression profiling reveals off-target gene regulation by RNAi. *Nat Biotechnol* **21:** 635–637.

Kent WJ, Sugnet CW, Furey TS, Roskin KM, Pringle TH, Zahler AM, Haussler D. 2002. The human genome browser at UCSC. *Genome Res* **12:** 996–1006.

Kleinstiver BP, Prew MS, Tsai SQ, Topkar VV, Nguyen NT, Zheng Z, Gonzales AP, Li Z, Peterson RT, Yeh JR, et al. 2015. Engineered CRISPR-Cas9 nucleases with altered PAM specificities. *Nature* **523:** 481–485.

Koike-Yusa H, Li Y, Tan EP, Velasco-Herrera Mdel C, Yusa K. 2014. Genome-wide recessive genetic screening in mammalian cells with a lentiviral CRISPR-guide RNA library. *Nat Biotechnol* **32:** 267–273.

Konermann S, Brigham MD, Trevino AE, Joung J, Abudayyeh OO, Barcena C, Hsu PD, Habib N, Gootenberg JS, Nishimasu H, et al. 2015. Genome-scale transcriptional activation by an engineered CRISPR-Cas9 complex. *Nature* **517:** 583–588.

Lin S, Ewen-Campen B, Ni X, Housden BE, Perrimon N. 2015. *In vivo* transcriptional activation using CRISPR-Cas9 in *Drosophila. Genetics* **201:** 433–442.

Liu H, Wei Z, Dominguez A, Li Y, Wang X, Qi LS. 2015. CRISPR-ERA: A comprehensive disign tool for CRISPR-mediated gene editing, repression, and activation. *Bioinformatics* **31:** 3676–3678.

Lowder LG, Zhang D, Baltes NJ, Paul JW III, Tang X, Zheng X, Voytas DF, Hsieh TF, Zhang Y, Qi Y. 2015. A CRISPR/Cas9 toolbox for multiplexed plant genome editing and transcriptional regulation. *Plant Physiol* **169:** 971–985.

Nishimasu H, Cong L, Yan WX, Ran FA, Zetsche B, Li Y, Kurabayashi A, Ishitani R, Zhang F, Nureki O. 2015. Crystal structure of *Staphylococcus aureus* Cas9. *Cell* **162:** 1113–1126.

Paul CP, Good PD, Winer I, Engelke DR. 2002. Effective expression of small interfering RNA in human cells. *Nat Biotechnol* **20:** 505–508.

Qi LS, Larson MH, Gilbert LA, Doudna JA, Weissman JS, Arkin AP, Lim WA. 2013. Repurposing CRISPR as an RNA-guided platform for sequence-specific control of gene expression. *Cell* **152:** 1173–1183.

Ran FA, Cong L, Yan WX, Scott DA, Gootenberg JS, Kriz AJ, Zetsche B, Shalem O, Wu X, Makarova KS, et al. 2015. In vivo genome editing using *Staphylococcus aureus* Cas9. *Nature* **520:** 186–191.

Shalem O, Sanjana NE, Hartenian E, Shi X, Scott DA, Mikkelsen TS, Heckl D, Ebert BL, Root DE, Doench JG, et al. 2014. Genome-scale CRISPR-Cas9 knockout screening in human cells. *Science* **343:** 84–87.

Sigoillot FD, Lyman S, Huckins JF, Adamson B, Chung E, Quattrochi B, King RW. 2012. A bioinformatics method identifies prominent off-targeted transcripts in RNAi screens. *Nat Methods* **9:** 363–366.

Tanenbaum ME, Gilbert LA, Qi LS, Weissman JS, Vale RD. 2014. A protein-tagging system for signal amplification in gene expression and fluorescence imaging. *Cell* **159:** 635–646.

Wang T, Wei JJ, Sabatini DM, Lander ES. 2014. Genetic screens in human cells using the CRISPR-Cas9 system. *Science* **343:** 80–84.

Zalatan JG, Lee ME, Almeida R, Gilbert LA, Whitehead EH, La Russa M, Tsai JC, Weissman JS, Dueber JE, Qi LS, et al. 2015. Engineering complex synthetic transcriptional programs with CRISPR RNA scaffolds. *Cell* **160:** 339–350.

Cite this protocol as *Cold Spring Harb Protoc*; doi:10.1101/pdb.prot090175

APPENDIX

General Safety and Hazardous Material Information

This manual should be used by laboratory personnel with experience in laboratory and chemical safety or students under the supervision of such trained personnel. The procedures, chemicals, and equipment referenced in this manual are hazardous and can cause serious injury unless performed, handled, and used with care and in a manner consistent with safe laboratory practices. Students and researchers using the procedures in this manual do so at their own risk. It is essential for your safety that you consult the appropriate Material Safety Data Sheets, the manufacturers' manuals accompanying equipment, and your institution's Environmental Health and Safety Office, as well as the General Safety and Disposal Cautions in this appendix for proper handling of hazardous materials in this manual. Cold Spring Harbor Laboratory makes no representations or warranties with respect to the material set forth in this manual and has no liability in connection with the use of these materials.

All registered trademarks, trade names, and brand names mentioned in this book are the property of the respective owners. Readers should please consult individual manufacturers and other resources for current and specific product information.

Users should always consult individual manufacturers, the manufacturers' safety guidelines and other resources, including local safety offices, for current and specific product information and for guidance regarding the use and disposal of hazardous materials.

PRIMARY SAFETY INFORMATION RESOURCES FOR LABORATORY PERSONNEL

Institutional Safety Office. The best source of toxicity, hazard, storage, and disposal information is your institutional safety office, which maintains and makes available the most current information. Always consult this office for proper use and disposal procedures.

Post the phone numbers for your local safety office, security office, poison control center, and laboratory emergency personnel in an obvious place in your laboratory.

Material Safety Data Sheets (MSDSs). The Occupational Safety and Health Administration (OSHA) requires that MSDSs accompany all hazardous products that are shipped. These data sheets contain detailed safety information. MSDSs should be filed in the laboratory in a central location as a reference guide.

GENERAL SAFETY AND DISPOSAL CAUTIONS

The guidance offered here is intended to be generally applicable. However, proper waste disposal procedures vary among institutions; therefore, always consult your local safety office for specific instructions. All chemically constituted waste must be disposed of in a suitable container clearly labeled with the type of material it contains and the date the waste was initiated.

It is essential for laboratory workers to be familiar with the potential hazards of materials used in laboratory experiments and to follow recommended procedures for their use, handling, storage, and disposal.

The following general cautions should always be observed.

- **Before beginning the procedure,** become completely familiar with the properties of substances to be used.
- **The absence of a warning** does not necessarily mean that the material is safe, because information may not always be complete or available.
- **If exposed** to toxic substances, contact your local safety office immediately for instructions.
- **Use proper disposal procedures** for all chemical, biological, and radioactive waste.
- **For specific guidelines on appropriate gloves to use,** consult your local safety office.
- **Handle concentrated acids and bases** with great care. Wear goggles and appropriate gloves. A face shield should be worn when handling large quantities.

  Do not mix strong acids with organic solvents because they may react. Sulfuric acid and nitric acid especially may react highly exothermically and cause fires and explosions.

  Do not mix strong bases with halogenated solvents because they may form reactive carbenes that can lead to explosions.
- **Handle and store pressurized gas containers** with caution because they may contain flammable, toxic, or corrosive gases; asphyxiants; or oxidizers. For proper procedures, consult the Material Safety Data Sheet that is required to be provided by your vendor.
- **Never pipette** solutions using mouth suction. This method is not sterile and can be dangerous. Always use a pipette aid or bulb.
- **Keep halogenated and nonhalogenated** solvents separately (e.g., mixing chloroform and acetone can cause unexpected reactions in the presence of bases). Halogenated solvents are organic solvents such as chloroform, dichloromethane, trichlorotrifluoroethane, and dichloroethane. Nonhalogenated solvents include pentane, heptane, ethanol, methanol, benzene, toluene, *N,N*-dimethylformamide (DMF), dimethylsulfoxide (DMSO), and acetonitrile.
- **Laser radiation**, visible or invisible, can cause severe damage to the eyes and skin. Take proper precautions to prevent exposure to direct and reflected beams. Always follow the manufacturer's safety guidelines and consult your local safety office. See caution below for more detailed information.
- **Flash lamps**, because of their light intensity, can be harmful to the eyes. They also may explode on occasion. Wear appropriate eye protection and follow the manufacturer's guidelines.
- **Photographic fixatives, developers, and photoresists** also contain chemicals that can be harmful. Handle them with care and follow the manufacturer's directions.
- **Power supplies and electrophoresis equipment** pose serious fire hazard and electrical shock hazards if not used properly.
- **Microwave ovens and autoclaves** in the laboratory require certain precautions. Accidents have occurred involving their use (e.g., when melting agar or Bacto Agar stored in bottles or when sterilizing). If the screw top is not completely removed and there is inadequate space for the steam to vent, the bottles can explode and cause severe injury when the containers are removed from the microwave or autoclave. Always completely remove bottle caps before microwaving or autoclaving. An alternative method for routine agarose gels that do not require sterile agar is to weigh out the agar and place the solution in a flask.
- **Ultrasonicators** use high-frequency sound waves (16–100 kHz) for cell disruption and other purposes. This "ultrasound," conducted through air, does not pose a direct hazard to humans, but the associated high volumes of audible sound can cause a variety of effects, including headache, nausea, and tinnitus. Direct contact of the body with high-intensity ultrasound (not medical

imaging equipment) should be avoided. Use appropriate ear protection and display signs on the door(s) of laboratories where the units are used.

- **Use extreme caution when handling cutting devices,** such as microtome blades, scalpels, razor blades, or needles. Microtome blades are extremely sharp! Use care when sectioning. If unfamiliar with their use, have an experienced user demonstrate proper procedures. For proper disposal, use the "sharps" disposal container in your laboratory. Discard used needles *unshielded*, with the syringe still attached. This prevents injuries and possible infections when manipulating used needles because many accidents occur while trying to replace the needle shield. Injuries may also be caused by broken Pasteur pipettes, coverslips, or slides.
- **Procedures for the humane treatment of animals** must be observed at all times. Consult your local animal facility for guidelines. Animals, such as rats, are known to induce allergies that can increase in intensity with repeated exposure. Always wear a lab coat and gloves when handling these animals. If allergies to dander or saliva are known, wear a mask.

DISPOSAL OF LABORATORY WASTE

There are specific regulatory requirements for the disposal of all medical waste and biological samples mandated by the U.S. Environmental Protection Agency (see http://www.epa.gov/epawaste/hazard/tsd/index.htm) and regulated by the individual states and territories (see http://www.epa.gov/epawaste/wyl/stateprograms.htm). Medical and biological samples that require special handling and disposal are generally termed Medical Pathological Waste (MPW), and medical, veterinary, and biological facilities will have programs for the collection of MPW and its disposal. Restrictions on how radioactive waste can be disposed of as regulated by the U.S. Nuclear Regulatory Commission can be found in 10 CFR 20.2001, General requirements for waste disposal (see http://www.nrc.gov/reading-rm/doc-collections/cfr/part020/part020-2001.html) or the individual Agreement States. The preferred method for the disposal of radioactively contaminated MPW'is decay-in-storage (see http://www.nrc.gov/reading-rm/doc-collections/cfr/part035/part035-0092.html).

Waste and any materials contaminated with biohazardous materials must be decontaminated and disposed of as regulated medical waste. No harmful substances should be released into the environment in an uncontrolled manner. This includes all tissue samples, needles, syringes, scalpels, etc. Be sure to contact your institution's safety office concerning the proper practices associated with the handling and disposal of biohazardous waste.

Some basic rules are outlined below. For treatment of radioactive and biological waste, see sections on Radioactive Safety Procedures and Biological Safety Procedures.

- In practice, only **neutral aqueous solutions** without heavy metal ions and without organic solvents can be poured down the drain (e.g., most buffers). Acid and basic aqueous solutions need to be neutralized cautiously before their disposal by this method.
- For proper disposal of **strong acids and bases,** dilute them by placing the acid or base onto ice and neutralize them. Do not pour water into them. If the solution does not contain any other toxic compound, the salts can be flushed down the drain.
- For disposal of **other liquid waste,** similar chemicals can be collected and disposed of together, whereas chemically different wastes should be collected separately. This avoids chemical reactions between components of the mixture (see above). Collect at least inorganic aqueous waste, nonhalogenated solvents, and halogenated solvents separately.
- Waste **from photo processing and automatic developers** should be collected separately to recycle the silver traces found in it.

RADIOACTIVE SAFETY PROCEDURES

In the United States and other countries, the access to radioactive substances is strictly controlled. You may be required to become a registered user (e.g., by attending a mandatory seminar and receiving a personal dosimeter). A convenient calculator to perform routine radioactivity calculations can also be found at http://www.graphpad.com/quickcalcs/ChemMenu.cfm.

If you have never worked with radioactivity before, follow the steps below.

- ***Try to avoid it!*** Many experiments that are traditionally performed with the help of radioactivity can now be done using alternatives based on fluorescence or chemiluminescence and colorimetric assays, including, for example, DNA sequencing, Southern and northern blots, and protein kinase assays. However, in other cases (e.g., metabolic labeling of cells), use of radioactivity cannot be avoided.
- **Be informed.** While planning an experiment that involves the use of radioactivity, include the physicochemical properties of the isotope (half-life, emission type, and energy), the chemical form of the radioactivity, its radioactive concentration (specific activity), total amount, and its chemical concentration. Order and use only as much as is really needed.
- **Familiarize yourself** with the designated working area. Perform a mental and practical dry run (replacing radioactivity with a colored solution) to make sure that all equipment needed is available and to get used to working behind a shield. Handle your samples as if sterility would be required to avoid contamination.
- **Always wear appropriate gloves**, lab coat, and safety goggles when handling radioactive material.
- **Check the work area** for contamination before, during, and after your experiment (including your lab coat, hands, and shoes).
- **Localize your radioactivity.** Avoid formation of aerosols or contamination of large volumes of buffers.
- **Liquid scintillation cocktails** are often used to quantitate radioactivity. They contain organic solvents and small amounts of organic compounds. Try to avoid contact with the skin. After use, they should be regarded as radioactive waste; the filled vials are usually collected in designated containers, separate from other (aqueous) liquid radioactive waste.
- **Dispose of radioactive waste** only into designated, shielded containers (separated by isotope, physical form [dry/liquid], and chemical form [aqueous/organic solvent phase]). Always consult your safety office for further guidance in the appropriate disposal of radioactive materials.
- Among the experiments requiring **special precautions** are those that use [^{35}S]methionine and ^{125}I, because of the dangers of airborne radioactivity. [^{35}S]methionine decomposes during storage into sulfoxide gases, which are released when the vial is opened. The isotope ^{125}I accumulates in the thyroid and is a potential health hazard. ^{125}I is used for the preparation of Bolton–Hunter reagent to radioiodinate proteins. Consult your local safety office for further guidance in the appropriate use and disposal of these radioactive materials before initiating any experiments. Wear appropriate gloves when handling potentially volatile radioactive substances, and work only in a radioiodine fume hood.

BIOLOGICAL SAFETY PROCEDURES

Biological safety fulfills three purposes: to avoid contamination of your biological sample with other species; to avoid exposure of the researcher to the sample; and to avoid release of living material into the environment. Biological safety begins with the receipt of the living sample; continues with its storage, handling, and propagation; and ends only with the proper disposal of all contaminated materials. A catalog of operations known as "sterile handling" is usually employed in manipulating living matter. However, the actual manner of treatment largely depends on the actual sample, which can be quite diverse: *Escherichia coli* and other bacterial strains, yeasts, tissues of animal or plant origin, cultures of mammalian cells, or even derivatives from human blood are routinely handled in a biological laboratory. Two of these, bacteria and human blood products, are discussed in more detail below.

The Department of Health, Education, and Welfare (HEW) has classified various bacteria into different categories with regard to shipping requirements (see Sanderson and Zeigler 1991). Non-pathogenic strains of *E.coli* (such as K12) and *Bacillus subtilis* are in Class 1 and are considered to present no or minimal hazard under normal shipping conditions. However, *Salmonella*, *Haemophilus*, and certain strains of *Streptomyces* and *Pseudomonas* are in Class 2. Class 2 bacteria are "[a]gents of ordinary potential hazard: agents which produce disease of varying degrees of severity... but which are contained by ordinary laboratory techniques." Contact your institution's safety office concerning shipping biological material.

Human blood, blood products, and tissues may contain occult infectious materials such as hepatitis B virus and human immunodeficiency virus (HIV) that may result in laboratory-acquired infections. Investigators working with lymphoblast cell lines transformed by Epstein–Barr virus (EBV) are also at risk of EBV infection. Any human blood, blood products, or tissues should be considered a biohazard and should be handled accordingly until proved otherwise. Wear appropriate disposable gloves, use mechanical pipetting devices, work in a biological safety cabinet, protect against the possibility of aerosol generation, and disinfect all waste materials before disposal. Autoclave contaminated plasticware before disposal; autoclave contaminated liquids or treat with bleach (10% [v/v] final concentration) for at least 30 minutes before disposal (this is valid also for used bacterial media).

Always consult your local institutional safety officer for specific handling and disposal procedures of your samples. Further information can be found in the Frequently Asked Questions of the ATCC homepage (http://www.atcc.org) and is also available from the National Institute of Environmental Health and Human Services, Biological Safety (http://www.niehs.nih.gov/about/stewardship).

GENERAL PROPERTIES OF COMMON HAZARDOUS CHEMICALS

The hazardous materials list can be summarized in the following categories.

- **Inorganic acids**, such as hydrochloric, sulfuric, nitric, or phosphoric, are colorless liquids with stinging vapors. Avoid spills on skin or clothing. Spills should be diluted with large amounts of water. The concentrated forms of these acids can destroy paper, textiles, and skin and cause serious injury to the eyes.
- **Inorganic bases**, such as sodium hydroxide, are white solids that dissolve in water and under heat development. Concentrated solutions will slowly dissolve skin and even fingernails.
- **Salts of heavy metals** are usually colored, powdered solids that dissolve in water. Many of them are potent enzyme inhibitors and therefore toxic to humans and the environment (e.g., fish and algae).
- Most **organic solvents** are flammable volatile liquids. Avoid breathing the vapors, which can cause nausea or dizziness. Also avoid skin contact.
- **Other organic compounds** including organosulfur compounds, such as mercaptoethanol or organic amines, can have very unpleasant odors. Others are highly reactive and should be handled with appropriate care.
- If improperly handled, **dyes and their solutions** can stain not only your sample but also your skin and clothing. Some are also mutagenic (e.g., ethidium bromide), carcinogenic, and toxic.

- **Nearly all names ending with "ase"** (e.g., catalase, β-glucuronidase, or zymolyase) refer to enzymes. There are also other enzymes with nonsystematic names such as pepsin. Many of them are provided by manufacturers in preparations containing buffering substances, etc. Be aware of the individual properties of materials contained in these substances.
- **Toxic compounds** are often used to manipulate cells. They can be dangerous and should be handled appropriately.
- Be aware that several of the compounds listed have not been thoroughly studied with respect to their toxicological properties. Handle each chemical with appropriate respect. Although the toxic effects of a compound can be quantified (e.g., LD_{50} values), this is not possible for carcinogens or mutagens where one single exposure can have an effect. Also realize that dangers related to a given compound may also depend on its physical state (fine powder vs. large crystals/diethyl ether vs. glycerol/dry ice vs. carbon dioxide under pressure in a gas bomb). Anticipate under which circumstances during an experiment exposure is most likely to occur and how best to protect yourself and your environment.

Cold Spring Harbor Laboratory Press (CSHLP) has used its best efforts in collecting and preparing the material contained herein but does not assume, and hereby disclaims, any liability for any loss or damage caused by errors and omissions in the publication, whether such errors and omissions result from negligence, accident, or any other cause. CSHLP does not assume responsibility for the user's failure to consult more complete information regarding the hazardous substances listed in this publication.

REFERENCE

Sanderson KE, Zeigler DR. 1991. Storing, shipping, and maintaining records on bacterial strains. *Methods Enzymol* **204:** 248–264.

WWW RESOURCES

ATCC Home page http://www.atcc.org

ATCC, for Sample Handling (in Frequently Asked Questions) http://www.atcc.org/CulturesandProducts/TechnicalSupport/FrequentlyAskedQuestions/tabid/469/Default.aspx

GraphPad Software, Radioactivity Calculations http://www.graphpad.com/quickcalcs/ChemMenu.cfm

National Institute of Environmental Health and Human Services, Biological Safety (NIEHS) http://www.niehs.nih.gov/about/stewardship

U.S. Environmental Protection Agency (EPA), Federal waste disposal regulations, Laboratory http://www.epa.gov/epawaste/hazard/tsd/index.htm

U.S. Environmental Protection Agency (EPA), Individual States and Territories http://www.epa.gov/epawaste/wyl/stateprograms.htm

U.S. Nuclear Regulatory Commission (NRC), Medical Pathological Radioactively Contaminated Waste (Decay-in-Storage) http://www.nrc.gov/reading-rm/doc-collections/cfr/part035/part035-0092.html

U.S. Nuclear Regulatory Commission (NRC), Radioactive Waste Disposal Regulations: General Requirements http://www.nrc.gov/reading-rm/doc-collections/cfr/part020/part020-2001.html

Index